T0257210

ANNALS OF MATHEMATICS STUDIES
NUMBER 11

INTRODUCTION TO NON-LINEAR MECHANICS

BY

N. KRYLOFF and N. BOGOLIUBOFF

A free translation by Solomon Lefschetz
of excerpts from two Russian monographs

PRINCETON
PRINCETON UNIVERSITY PRESS
LONDON: HUMPHREY MILFORD
OXFORD UNIVERSITY PRESS
1949

(Reprinted 1947 with errata on page 106)

INTRODUCTION

During the last decade a number of Soviet scientists have investigated so-called non-linear mechanics, and among the most active are certainly to be found Kryloff and Bogoliuboff. An extensive bibliography of their contributions to the subject will be found at the end. A cursory reference to it will quickly disclose the fact that in one way or another their work is but poorly accessible to the American scientific and technical public. The present monograph is essentially a very condensed English version of their most extensive paper (No. 32 of the Bibliography, in Russian) except for the last chapter which is practically a small extract of their most mathematical production on the subject (No. 16 of the Bibliography, in Russian).

Kryloff and Bogoliuboff consider primarily equations of the form

$$\frac{d^2x}{dt^2} + \omega^2 x = \varepsilon f(t, x, \frac{dx}{dt}, \varepsilon)$$

where ε is a small positive quantity and f is a power series in ε, whose coefficients are polynomials in x, $\frac{dx}{dt}$, sin t, cos t. As a matter of fact, generally f contains neither ε nor t. Similar equations are well known in astronomy and have been the object of systematic investigation by Linstedt, Gyldén, Liapounoff and, above all by Poincaré. In a general sense, one may say that the same methods are applied by Kryloff and

Bogoliuboff. However, the applications which they have in view are quite different, being chiefly in Engineering, Technology, or Physics, notably electrical circuit theory. The solutions are approximated by the first n terms of certain asymptotic representations; the first two terms usually suffice and yield what the authors call the "refined first approximation" which they discuss at length.

The method of linearization described in Chapters V and VI, frequently enables one to by-pass the differential equation and proceed directly from the physical problem to the approximate solutions. That the general information obtained from the approximations gives important indications regarding the behaviour of the solution itself, is shown in the monograph, (No. 16 of the Bibliography) of which the extract given in Chapter IX will yield a few indications.

Messrs. Kryloff and Bogoliuboff deserve much credit for the bold way in which they have carried out their work and for the numerous applications which they have outlined. It is believed that the present monograph will provide a fair picture of what they have accomplished.

S. Lefschetz
Princeton, N. J.
November 20, 1942.

TABLE OF CONTENTS

Chapter I. Some Non-Linear Oscillatory Systems	1
Chapter II. Elementary Theory of the First Approximation	8
Chapter III. Refinement of the First Approximation	28
Chapter IV. Construction of the Higher Approximations	40
Chapter V. Linearization	55
Chapter VI. Application of Symbolic Methods To Linearization	63
Chapter VII. Multiply Periodic Systems	73
Chapter VIII. Influence of Periodic Disturbances	79
Chapter IX. Complements	87
Bibliography	100

I. SOME NON-LINEAR OSCILLATORY SYSTEMS.

1. In the present section we will discuss a few non-linear oscillatory systems and derive the corresponding differential equations. These equations will serve later as illustrations for the methods of approximation introduced in the sequel.

2. We begin with some conservative (non-dissipative) systems.

(2.1) Oscillating shaft. Consider a shaft composed (ideally) of two revolving masses joined by a non-linear elastic connection. Let θ_1, θ_2 be the moments of inertia of the revolving masses, and θ_1, θ_2 their angles of rotation. Let further $M = c(\theta_1 - \theta_2)$ be the angular momentum of the elastic connection represented as a function of the angle of rotation $\theta = \theta_1 - \theta_2$. The equations of motion for each of the two masses are

$$J_1 \frac{d^2\theta_1}{dt^2} + c(\theta_1 - \theta_2) = 0,$$

$$J_2 \frac{d^2\theta_2}{dt^2} - c(\theta_1 - \theta_2) = 0.$$

Hence the equation governing the oscillations is

$$\frac{d^2\theta}{dt^2} + \frac{J_1 + J_2}{J_1 \; J_2} c(\theta) = 0. \tag{2.2}$$

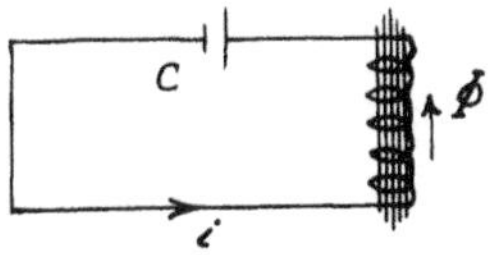

Fig. 1

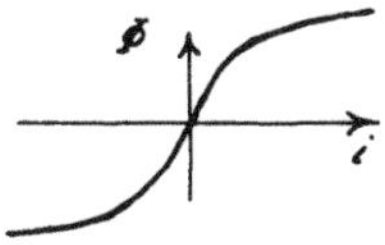

Fig. 2

In this relation the function $c(\theta)$ is usually given graphically and may have the most diverse form.

(2.3) Electrical circuit without resistance. Consider an electrical oscillating circuit (Fig. 1) containing an iron core. Let Φ denote the magnetic flux, i the line current, C the capacity. We then have

$$(2.4)\quad \frac{d\Phi}{dt} + \frac{1}{C}\int^{t} i\,dt = 0.$$

The relation between Φ and i is shown in Fig. 2. With sufficient accuracy and within certain limits, one may represent this relation analytically, for instance as:

$$(2.5)\quad i = A\Phi + B\Phi^3$$

We have then for Φ the differential equation

$$(2.6)\quad \frac{d^2\Phi}{dt^2} + \frac{A\Phi+B\Phi^3}{C} = 0.$$

3. In the examples of oscillating systems that we have examined so far, we have not taken into consideration friction which causes dissipation of the oscillations of the system.

Generally speaking the laws of mecnanical friction have been but little investigated. In practice, one chiefly assumes one of the following three:

a) The force of friction is proportional to the velocity (oscillations in the atmosphere).

b) The force of friction is proportional to the square of the velocity (for oscillations in a liquid).

c) Coulomb's law: The force of friction is constant in magnitude but depends upon the velocity and its direction is opposite the velocity (for example in slipping of surfaces upon one another).

(3.1) Pendulum freely oscillating in the atmosphere. If we assume that friction is proportional to the velocity, the equation of oscillations will be

(3.2) $$\frac{d^2\theta}{dt^2} + \lambda\frac{d\theta}{dt} + \frac{g}{l}\sin\theta = 0$$

where λ is a proportionality coefficient, called friction coefficient.

(3.3) Electrical circuit with resistance. We suppose that the circuit contains an iron core, an ohmic resistance and a capacity (Fig. 3). Let Φ be the flux, i the current, R the ohmic resistance, C the capacity. We will have this time

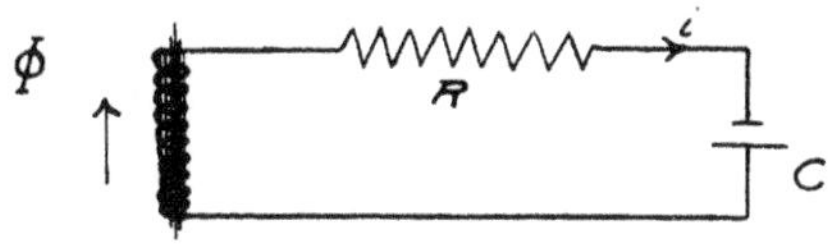

Fig. 3

$$\frac{d\Phi}{dt} + Ri + \frac{1}{C}\int^{t} i\,dt = 0$$

and hence assuming that (2.5) holds:

(3.4) $$\frac{d^2\Phi}{dt^2} + R(A+3B\Phi^2)\frac{d\Phi}{dt} + \frac{A\Phi+B\Phi^3}{C} = 0.$$

4. Up to the present, we have considered oscillating systems with or without dissipation (friction). Since in practice dissipation is always present in some form in oscillating systems, the oscillations will fail to die down only if the system contains some source of energy which may compensate for the loss of energy due to dissipation. This condition may be fulfilled in two ways. First, the force acting upon the oscillating body (due to its connection with the source of energy), may possess a definite periodicity. The simplest example of oscillations of this type, said to be forced, is found in the vibrations of linear

systems subjected to a harmonic disturbance:

$$(4.1)\quad m\frac{d^2x}{dt^2} + \lambda\frac{dx}{dt} + kx = F\sin\alpha t$$

where m is the mass, x the displacement, λ the dissipation coefficient, k the spring constant, F the amplitude of the exterior force, α the frequency of the disturbance.

Second, the source of energy itself may have no specific periodicity but its action upon the oscillating body appears to introduce into the system a negative dissipation which may compensate for the normal positive dissipation caused by the dissipative forces. Oscillations of this last type, called auto-oscillations, are quite wide-spread and have great importance in Physics and Technology.

To obtain some idea of the manner in which auto-oscillations arise, we will examine a system with one degree of freedom.

If the oscillations are of rather small amplitude we may write down the customary linear equation:

$$(4.2)\quad m\frac{d^2x}{dt^2} + \lambda\frac{dx}{dt} + kx = 0.$$

As is well known the general solution will be

$$x = ae^{-\delta t}\cos(\omega t+\phi)$$

where a, ϕ are arbitrary constants,

$$\delta = \frac{\lambda}{2m}, \quad \omega^2 = \frac{k}{m} - \left(\frac{\lambda}{m}\right)^2$$

Hence if $\lambda > 0$, then the amplitude of the small oscillations $ae^{-\delta t}$ will die down according to an exponential law. If on the contrary $\lambda < 0$ then the small oscillations will expand and the amplitudes will increase exponentially.

Since for physical reasons the amplitudes cannot increase indefinitely, we must suppose that from a certain moment the dissipation coefficient changes its sign and becomes positive. This fact may be reflected in the differential equation of the oscillations, for instance, by replacing the constant coefficient λ by a variable one:

$$\lambda = -A + B \left(\frac{dx}{dt}\right)^2$$

where $A > 0$, $B > 0$. We thus obtain a differential equation due to Rayleigh:

$$(4.3) \qquad m \frac{d^2x}{dt^2} + \left(-A + B\left(\frac{dx}{dt}\right)^2\right) \frac{dx}{dt} + kx = 0.$$

This equation shows in particular that the dissipation is negative for small absolute values of $\frac{dx}{dt}$ and positive when its absolute values are large.

Thus, small oscillations will expand and large oscillations will die down.

The importance of (4.3) for self-oscillatory systems was already brought out by Rayleigh in his paper: <u>On maintained vibrations</u> (Phil. Mag. S. 5, vol. 15, 1883).

Another important equation as regards self-oscillatory systems, repeatedly investigated by van der Pol and going by his name, is

$$(4.4) \qquad \frac{d^2x}{dt^2} - \varepsilon(1-x^2) \frac{dx}{dt} + x = 0.$$

It may be deduced from (4.3) by making the change of variables:

$$t\sqrt{\frac{k}{m}} \longrightarrow t, \quad \frac{dx}{dt}\sqrt{\frac{3Bk}{Am}} \longrightarrow x$$

and setting $\frac{A}{\sqrt{km}} = \varepsilon$.

5. We will now consider some self-oscillatory systems.

(5.1) Electronic generator. We refer to Fig. 4 for the various designations of currents (written i with subscripts), voltages (written V, E with subscripts), etc.

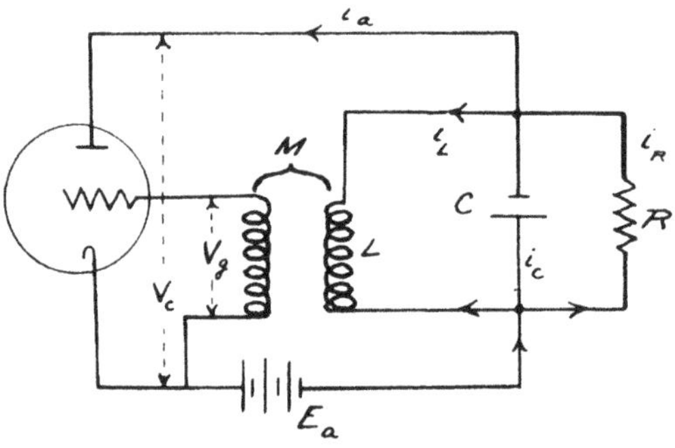

Fig. 4

Neglecting the grid current, we have clearly:

$$(5.2)\quad \begin{cases} L\dfrac{di_L}{dt} = \dfrac{1}{C}\displaystyle\int^{t} i_c dt = Ri_R = E_a - V_a, \\ M\dfrac{di_L}{dt} = V_g \quad , \; i_a = i_L + i_C + i_R. \end{cases}$$

From (5.2) we find:

$$(5.3)\qquad LC\frac{d^2 i_L}{dt^2} + \frac{L}{R}\frac{di_L}{dt} + i_L = i_a.$$

As we know, however, from the theory of electronic lamps, the anode current is a definite function of the so called directing potential $u = V_g + DV_a$

(5.4) $i_a = f(u) = f(V_g + DV_a)$,

where D is a constant factor, the conductance of the lamp.

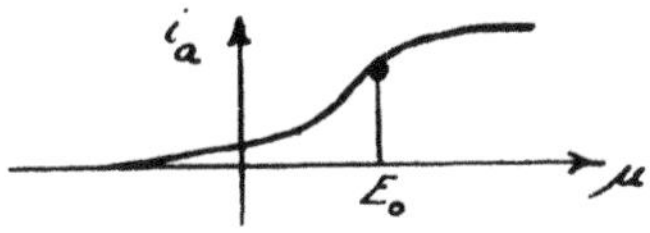

Fig. 5

In practice D is small relatively to unity. A typical curve representing the relation (5.4), the so-called characteristic of the lamp is shown in Fig. 5.

Substituting from (5.4) into (5.3) and in view of (5.2) we find

$$(5.5)\quad LC\frac{d^2i_L}{dt^2} + \frac{L}{R}\frac{di_L}{dt} + i_L = f(DE_a+(M-LD)\frac{di_L}{dt}).$$

Consider now the following quantities:

$$E_o = DE_a,\ V = (M-LD)\frac{di_L}{dt}.$$

Since the directing potential is

$$DE_a + (M-LD)\frac{di_L}{dt} = E_o + V,$$

V will clearly be the variable part of this potential induced by the vibrations of the current in the oscillating circuit, and E_o the constant part induced by a source of constant current (for instance a battery). In view of this, let us apply to both sides of (5.5) the operation

$$(M-LD)\frac{d}{dt}.$$

We then obtain for the unknown V a relationship of the form

$$(5.6)\quad LC\frac{d^2V}{dt^2} + V + \{\frac{L}{R} - (M-LD)f'(E_o + V)\}\frac{dV}{dt} = 0.$$

If we choose E_0 such that it is the abscissa of the inflexion of the characteristic in Fig. 5, and neglect terms of order > 3 in the MacLaurin expansion of $f(E_0 + V)$, then with suitable choice of a time unit (dimensionless time), (5.6) may be reduced to van der Pol's equation (4.4). Similar considerations lead to a Rayleigh differential equation for i_L.

(5.7) It is hardly necessary to observe that if a harmonic disturbance is superimposed upon any one of the preceding systems, there is obtained an equation for forced oscillations. Thus we may have in relation to a van der Pol system an equation

$$\frac{d^2x}{dt^2} - \varepsilon (1-x^2)\frac{dx}{dt} + x = F \sin \alpha t, \tag{5.8}$$

and likewise for the other systems.

II. ELEMENTARY THEORY OF THE FIRST APPROXIMATION

6. All the examples discussed in the preceding chapter lead to equations of the form

$$\frac{d^2x}{dt^2} + F(x,\frac{dx}{dt},t) = 0. \tag{6.1}$$

We propose to investigate more particularly the so called quasi-harmonic case, where there are oscillations near the sinusoidal:

$$x = a \sin (\nu t+\phi),$$

that is to say when we may write

$$F(x,\frac{dx}{dt},t) = \nu^2 x+\varepsilon f(x,\frac{dx}{dt},t), \tag{6.2}$$

where ε is a parameter characterizing the smallness of the deviation of F from $\nu^2 x$. Until further notice we assume F, and hence also f, free from the explicit variable t. The basic differential equation will thus be

$$\frac{d^2x}{dt^2} + \nu^2 x + \varepsilon f(x,\frac{dx}{dt}) = 0, \tag{6.3}$$

and this is the equation which we shall investigate. If we endeavor to solve this equation by the usual methods of approximation, notably by the method of Poisson, we encounter a classical difficulty which baffled the astronomers of the eighteenth century, namely the presence of so-called secular terms, or terms of the form t×(a trigonometric function). In the same spirit as the astronomers did in their day, we shall endeavor to find methods of approximation which yield results free from secular terms.

In the present chapter we shall describe a very intuitive method enabling us to construct an approximate solution which will be free from secular terms.

7. We first observe that for $\varepsilon = 0$, (6.3) has the solution

$$x = a \sin (\nu t+\phi), \tag{7.1}$$

$$\frac{dx}{dt} = a\nu \cos (\nu t+\phi), \tag{7.2}$$

where the amplitude a and the phase ϕ are constant. For convenience the term "frequency" will designate ν rather than the customary $2\pi\nu$.

Consider a, ϕ, as new unknown functions of the time which are to be determined so that (7.1) becomes a solution of (6.3). We must have first

$$\frac{dx}{dt} = \frac{da}{dt} \sin (\nu t+\phi) + a\frac{d\phi}{dt} \cos (\nu t+\phi) + a\nu \cos(\nu t+\phi). \tag{7.3}$$

Hence if we wish to preserve (7.2), or we may say if we impose (7.2), then

(7.4) $$\frac{da}{dt}\sin(\nu t+\phi) + a\frac{d\phi}{dt}\cos(\nu t+\phi) = 0.$$

From these relations we deduce:

(7.5) $$\frac{d^2x}{dt^2} = \frac{da}{dt}\cos(\nu t+\phi) - \nu a\frac{d\phi}{dt}\sin(\nu t+\phi)$$
$$-\nu^2 a\sin(\nu t+\phi),$$

and so finally from (6.3):

(7.6) $$\nu\frac{da}{dt}\cos(\nu t+\phi) - \nu a\frac{d\phi}{dt}\sin(\nu t+\phi)$$
$$= -\varepsilon f(a\sin(\nu t+\phi), a\nu\cos(\nu t+\phi)).$$

By combining with (7.4) there comes

(7.7) $$\frac{da}{dt} = -\frac{\varepsilon}{\nu}f(a\sin(\nu t+\phi), a\nu\cos(\nu t+\phi))\cos(\nu t+\phi),$$

(7.8) $$\frac{d\phi}{dt} = \frac{\varepsilon}{a\nu}f(a\sin(\nu t+\phi), a\nu\cos(\nu t+\phi))\sin(\nu t+\phi).$$

8. Thus instead of the single differential equation of the second order (6.3) in the unknown x, we have two differential equations of the first order in the two unknowns a, ϕ. Notice now that the right hand sides of (7.7), (7.8) admit with respect to t the period $T = \frac{2\pi}{\nu}$. Moreover $\frac{da}{dt}$, $\frac{d\phi}{dt}$ are proportional to the small perameter ε, so that a, ϕ will be slowly varying functions of the time during the period T, and as a first approximation we may, therefore, consider them as constant. On the strength of this observation, we will indicate at once a simple intuitive method for constructing an approximate solution of (7.7), (7.8). For this purpose consider the expressions

$f(a \sin \phi, a\nu \cos \phi) \cos \phi, f(a \sin \phi, a\nu \cos \phi) \sin \phi,$

and let us expand them in Fourier series. We find

$$(8.1)\quad \begin{cases} f(a \sin \phi,\ a\nu \cos \phi) \cos \phi = K_o(a) + \sum\limits_{n>o} (K_n(a) \\ \qquad\qquad \cos n\phi + L_n(a) \sin n\phi), \\ f(a \sin \phi, a\nu \cos \phi) \sin \phi = P_o(a) + \sum\limits_{n>o} (P_n(a) \\ \qquad\qquad \cos n\phi + Q_n(a) \sin n\phi. \end{cases}$$

The coefficients $P_n(a)$ . . . . , are calculated in the usual way. It will be sufficient to give the explicit expressions:

$$(8.2)\quad \begin{cases} K_o(a) = \frac{1}{2\pi} \int\limits_0^{2\pi} f(a \sin \phi,\ a\nu \cos \phi) \cos \phi d\phi, \\ P_o(a) = \frac{1}{2\pi} \int\limits_0^{2\pi} f(a \sin \phi,\ a\nu \cos \phi) \sin \phi d\phi. \end{cases}$$

Taking advantage of (8.1) we can represent (7.7), (7.8) in the following expanded forms:

$$(8.3)\quad \begin{cases} \frac{da}{dt} = - \frac{\varepsilon}{\nu} K_o(a) - \frac{\varepsilon}{\nu} \sum\limits_{n>o} (K_n(a) \cos n(\nu t + \phi) + L_n(a) \\ \qquad\qquad \sin n(\nu t + \phi)), \\ \frac{d\phi}{dt} = \frac{\varepsilon}{\nu a} P_o(a) + \frac{\varepsilon}{\nu a} \sum\limits_{n>o} (P_n(a) \cos n(\nu t + \phi) + Q_n(a) \\ \qquad\qquad \sin n(\nu t + \phi)). \end{cases}$$

Let us integrate these espressions in the interval $t,\ t + T$, within which we consider a, ϕ, as constant and equal to the values $a(t)$, $\phi(t)$.

We thus obtain:

$$(8.4)\qquad \begin{cases} \dfrac{a(t+T)-a(t)}{T} = -\dfrac{\varepsilon}{\nu}K_0(a(t)), \\ \dfrac{\phi(t+T)-\phi(t)}{T} = \dfrac{\varepsilon}{\sqrt{a}}P_0(a(t)). \end{cases}$$

Since T and the increments $a(t+T)-a(t)$, $\phi(t+T-\phi(t)$ are small, we replace in (8.4) the left sides by , $\frac{da}{dt}$, $\frac{d\phi}{dt}$, and thus arrive at the equations of the first approximation:

$$(8.5)\qquad \begin{cases} \dfrac{da}{dt} = -\dfrac{\varepsilon}{\nu}K_0(a) \\ \dfrac{d\phi}{dt} = \dfrac{\varepsilon}{\nu}P_0(a). \end{cases}$$

If we compare with the exact relations (8.3) we find that the equations of the first approximation are obtained from the exact equations by averaging the right hand sides with respect to the time. This process duly generalized in the obvious way will be described as the averaging principle.

It need not be said that the preceding reasoning cannot pretend to any sort of mathematical rigor. For this reason we shall examine in the next chapters the questions of the mathematical foundations of the averaging principle and likewise the question of forming the higher approximations.

9. Returning to (8.5), if we have a solution in a and ϕ and substitute it in (7.1), we obtain an approximate expression for x. If we choose in place of ϕ the unknown $\psi = \nu t + \phi$, then (8.5) yields

$$(9.1)\qquad \frac{d\psi}{dt} = \nu + \frac{\varepsilon}{\sqrt{a}}P_0(a).$$

Substituting in (8.5) and (9.1), in place of K_0,

P_0 their expressions from (8.2) we obtain explicitly

$$(9.2) \quad \frac{da}{dt} = \frac{-\varepsilon}{2\pi\nu} \int_0^{2\pi} f(\ \sin\phi, a\nu \cos\phi) \cos\phi \, d\phi,$$

$$(9.3) \quad \frac{d\psi}{dt} = \nu + \frac{\varepsilon}{2\pi a\nu} \int_0^{2\pi} f(a \sin\phi, a\nu \cos\phi) \sin\phi \, d\phi.$$

Thus the first approximation to the solution of (6.3) will be of the form

$$(9.4) \qquad x = a \sin\psi,$$

where the amplitude a and the full phase ψ are to be determined from (9.2), (9.3).

10. Suppose that F in (6.1) does not contain $\frac{dx}{dt}$, in which case f will likewise be free from it. Thus we will have

$$(10.1) \qquad f(x, \frac{dx}{dt}) = f(x),$$

and hence instead of (9.2), (9.3):

$$(10.2) \qquad \frac{da}{dt} = -\frac{\varepsilon}{2\pi\nu} \int_0^{2\pi} f(a \sin\phi) \cos\phi \, d\phi,$$

$$(10.3) \quad \frac{d\psi}{dt} = \omega(a) = \nu + \frac{\varepsilon}{2\pi a\nu} \int_0^{2\pi} f(a \sin\phi) \sin\phi \, d\phi.$$

If we set

$$\Phi(x) = \int_0^x f(x) dx$$

then

$$\int_0^{2\pi} f(a \sin\phi) \cos\phi \, d\phi = \frac{1}{a} \int^{2\pi} \frac{d\Phi(a \sin\phi)}{d\phi} = 0,$$

and hence

$$\frac{da}{dt} = 0.$$

Thus the amplitude of the oscillations is now constant $a = a_0$, and so instead of (10.3) we have

$$\psi = \omega(a)t+\theta,$$

where the phase θ is constant and equal to the initial value of ψ.

An approximate solution of (6.3) is then

$$x = a \sin (\omega(a)t+\theta). \tag{10.4}$$

We may say that here the nonlinear character of the equation has no other effect in the first approximation than to make the frequency depend upon the amplitude.

If we square both sides of (10.3) and retain only terms in ε we obtain

$$\omega^2(a) = \nu^2+\frac{\varepsilon}{a\pi}\int_0^{2\pi} f(a \sin \phi) \sin \phi d\phi. \tag{10.5}$$

Since $F(x) = \nu^2 x+\varepsilon f(x)$ we have finally:

$$\omega^2(a) = \frac{1}{\pi a}\int_0^{2\pi} F(a \sin \phi) \sin\phi \, d\phi. \tag{10.6}$$

Formula (10.6) has the considerable advantage that the function F enters into it directly and not merely through its nonlinear part as it does in (10.5).

11. We will now examine a certain number of examples.

(11.1) Example 1. Consider the equation of the pendulum reduced for small oscillations (say not exceeding 30°) to the form

$$\frac{d^2x}{dt^2} + \frac{g}{l}\left(x- \frac{x^3}{6}\right) = 0. \tag{11.2}$$

We have at once from (10.6):

$$\omega^2(a) = \frac{g}{l}\,\frac{1}{\pi a}\int_0^{2\pi}\left(a \sin\phi - \frac{a^3 \sin^3\phi}{6}\right)\sin\phi\, d\phi,$$

and so approximately

$$\omega^2(a) = \frac{g}{l}\left(1 - \frac{a^2}{8}\right). \tag{11.3}$$

As the amplitude increases the frequency decreases and hence the period increases also. This is likewise shown by the approximate formula

$$T = \frac{2\pi}{\omega} = 2\pi\sqrt{\frac{l}{g}}\left(\frac{1}{1-\frac{a^2}{16}}\right) = 2\pi\sqrt{\frac{l}{g}}\left(1+\frac{a^2}{16}\right). \tag{11.4}$$

To take a concrete example for $a = 30^\circ$ we find

$$T = 1.014 \times 2\pi\sqrt{\frac{l}{g}}.$$

(11.5) Example 2. Consider the differential equation (2.2) for the oscillations of a shaft. Here:

$$\omega^2(a) = \frac{J_1+J_2}{J_1 J_2}\,\frac{1}{\pi a}\int_0^{2\pi} c(a \sin\phi)\sin\phi\, d\phi. \tag{11.6}$$

To take a concrete case, suppose that $M = c(\theta)$ is represented by the graph of Fig. 6 , or more explicitly that

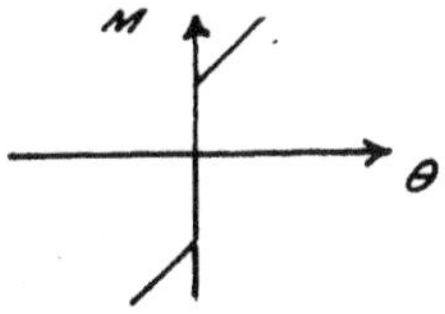

Fig. 6

$$c(\theta) = \begin{cases} h+k\theta, & \theta>0 \\ -h+k\theta, & \theta<0. \end{cases}$$

We find here

$$\int_0^{2\pi} c(a \sin \phi) \sin\phi \, d\phi = 4h + \pi ka$$

and so by (11.6):

$$(11.7) \qquad \omega^2(a) = \frac{J_1+J_2}{J_1 \, J_2} k(1+\frac{4h}{\pi ak}).$$

In order that this formula be applicable it will be clearly necessary that $\frac{h}{ak}$ be small. This quantity measures in a sense the deviation of M from linearity.

(11.8) Example 3. Take the case of the electrical circuit of (2.3) and related equation (2.5). Assuming $\frac{B\phi^2}{A}$ small, we find by (10.6):

$$(11.9) \qquad \omega^2(a) = \frac{A}{C}(1+\frac{3Ba^2}{4A})$$

from which follows approximately:

$$(11.10) \qquad \omega(a) = \sqrt{\frac{A}{C}}(1+\frac{3Ba^2}{8A}).$$

12. We will now examine some cases where F contains $\frac{dx}{dt}$.

(12.1) Example 4. Consider the equation of van der Pol (4.4) where the parameter ε is assumed small. Comparing with the basic equation (6.3) we have here:

$$\nu = 1, \; f(x,\frac{dx}{dt}) = -(1-x^2)\frac{dx}{dt}.$$

As a consequence we find

$$f(a \sin \phi, \, a\nu \cos \phi) = -a(1-a^2 \sin^2 \phi) \cos \phi$$

$$= a(\frac{a^2}{4} -1) \cos \phi - \frac{a^3}{4} \cos 3\phi,$$

and therefore

$$\frac{-1}{2\pi\nu}\int_0^{2\pi} f(a \sin \phi,\ a\nu \cos \phi) \cos \phi \, d\phi = \frac{a}{2}(1-\frac{a^2}{4}),$$

$$\frac{1}{2\pi\nu}\int_0^{2\pi} f(a \sin \phi,\ a\nu \cos \phi) \sin \phi \, d\phi = 0.$$

Thus referring to (9.2), (9.3), (9.4), we have in the first approximation

(12.1a) $$x = a \sin \psi$$

(12.2) $$\frac{da}{dt} = \frac{\varepsilon a}{2}(1-\frac{a^2}{4})$$

(12.3) $$\frac{d\psi}{dt} = 1.$$

From (12.3) we obtain $\psi = t + \theta$, where $\theta = \psi_0$.

Finally the first approximation is a harmonic oscillation

(12.4) $$x = a \sin (t+\theta)$$

with constant frequency whose amplitude varies in accordance with (12.2). By an elementary integration we obtain

$$a^2 = \frac{a_0^2 e^{\varepsilon t}}{1+\frac{1}{4}a_0^2(e^{\varepsilon t}-1)},$$

and hence finally

(12.5) $$a = \frac{a_0 e^{\frac{1}{2}\varepsilon t}}{\sqrt{1+\frac{1}{4}a_0^2(e^{\varepsilon t}-1)}}.$$

Substituting from (12.5) in (12.4) we obtain the explicit approximate expression for x:

$$(12.6) \qquad x = \frac{a_0 e^{\frac{1}{2}\varepsilon t}}{\sqrt{1+\frac{1}{4}a_0^2(e^{\varepsilon t}-1)}} \sin(t+\theta).$$

A trivial solution is $x = 0$ which corresponds to the static régime (without oscillations). It is not difficult to show, however, that this régime is not stable. Indeed however small the initial amplitude a_0 may be, it will grow monotonely tending to 2 as a limit. Thus the least disturbance will throw the system into an oscillation with growing amplitude.

From (12.5) we see also that if $a_0 = 2$, then $a = 2$ for all $t > 0$. This corresponds to the stationary régime

$$(12.7) \qquad x = 2\sin(t+\theta).$$

This "dynamical" régime is strongly stable, for whatever $a_0 (\neq 0)$, whether large or small, $a(t) \longrightarrow 2$ when $t \longrightarrow +\infty$. Thus an arbitrary oscillation will tend to the stationary oscillation (12.7).

The systems of the van der Pol type differ essentially from those of the conservative type with equations:

$$\frac{d^2x}{dt^2} + \nu^2 x + \varepsilon f(x) = 0.$$

Indeed in the conservative systems as we have seen there may occur steady oscillations of arbitrary amplitude whereas in the van der Pol system steady amplitudes are possible only for special values. Physically this is evident from the following considerations: since a conservative system neither dissipates nor creates energy, oscillations once started have no reason to die down or to grow and so their amplitudes remain fixed. On the contrary in a "self-exciting" system there is creation as well as dissipation of energy and so the amplitude may increase if the source of energy provides more energy that there is dissipated or conversely. There will thus arise a fixed amplitude only if the two processes compensate.

13. (13.1) Example 5. As our next example we will take Rayleigh's equation (4.3). Here the function f of (6.3) will be

$$f\left(\frac{dx}{dt}\right) = \left\{-A+B\left(\frac{dx}{dt}\right)^2\right\}\frac{dx}{dt}. \tag{13.2}$$

Hence we have

$$x = a \sin (\nu t+\theta), \tag{13.3}$$

as our first approximation, with

$$\nu = \sqrt{\frac{k}{m}}, \quad \theta = \text{const.},$$

and

$$\frac{da}{dt} = \frac{-1}{2\pi m\nu}\int_0^{2\pi} f(a\nu \cos \phi) \cos\phi \, d\phi. \tag{13.4}$$

However we find from (13.2)

$$f(a\nu \cos \phi) = -a\nu(A-\tfrac{3}{4}B^2a^2\nu^2) \cos \phi + \tfrac{1}{4}B(a\nu^3) \cos^3 \phi,$$

and so from (13.4):

$$\frac{da}{dt} = \frac{a}{2m}(A-\tfrac{3}{4}Ba^2\nu^2). \tag{13.5}$$

It follows from (13.5) that the trivial solution $a = 0$ will be unstable, since $A>0$, and so we have here a self-excited oscillation. The stationary amplitude satisfies

$$A - \tfrac{3}{4}Ba^2\nu^2 = 0,$$

which yields

$$(13.6) \qquad a = \frac{1}{\nu}\sqrt{\frac{4A}{3B}}.$$

Whatever the initial amplitude a_0 $(\neq 0)$ we have from (13.5):

$$a(t) \underset{t\to+\infty}{\longrightarrow} \frac{1}{\nu}\sqrt{\frac{4A}{3B}}\,.$$

Thus whatever the initial conditions the oscillation tends to a steady oscillation represented by

$$x = \frac{1}{\nu}\sqrt{\frac{4A}{3B}}\,\sin\,(\nu t+\theta).$$

If we desire to learn something not merely about the steady oscillations but about the imtermediary régime, we must integrate (13.5), which yields

$$a(t) = \frac{a_0 e^{\frac{A}{2m}t}}{\sqrt{1+\frac{3B\nu^2}{4A}a_0^2(e^{\frac{A}{m}t}-1)}}$$

14. (14.1) <u>Example 6</u>. As our next example we take an electrical circuit with constant capacity C and self-induction L, and containing a non-linear element N whose voltage-current characteristic is

$$(14.2) \qquad e = F(i).$$

The differential equation for i is

$$LC \frac{d^2 i}{dt^2} + CF'(i) \frac{di}{dt} + i = 0. \tag{14.3}$$

This equation is reduced to the form (6.3) by setting:

$$\nu^2 = \frac{1}{LC}, \varepsilon f(i, \frac{di}{dt}) = \frac{F'(i)}{L} \frac{di}{dt}.$$

In order to have a clear picture of the degree of smallness of the nonlinear element, it is convenient to introduce the dimensionless time $\tau = \frac{t}{\sqrt{LC}}$ which brings the equation to the form

$$\frac{d^2 i}{d\tau^2} + i + \frac{CF'(i)}{\sqrt{LC}} \frac{di}{d\tau} = 0.$$

This shows that the application of our results will require that the dimensionless quantity $\sqrt{\frac{C}{L}}\, F'(i)$ remain small relatively to unity.

We find here that the first approximation assumes the form

$$i = a \sin (\nu t + \phi), \ \phi = \text{const.},$$

where the amplitude a satisfies the equation

$$\frac{da}{dt} = - \frac{aR(a)}{2L} \tag{14.4}$$

with

$$R(a) = \frac{1}{\pi} \int_0^{2\pi} F'(a \sin t) \cos^2 t\, dt. \tag{14.5}$$

We see at once from (14.4) that if R(a) is always positive, $a(t) \to 0$ so that the oscillations die down. In this case steady oscillations with an amplitude other than zero are ruled out. Referring to (14.5) this will certainly occur whenever $F'(i) > 0$ for all i.

Thus if the characteristic $e = F(i)$ of the non-linear element does not have a falling part (where $F'(i) < 0$) then the system is dissipative: oscillations once started die down. If on the contrary there is a falling part in the characteristic then R(a) will be positive, at least for small values of a. In this case small amplitudes will increase and small oscillations expand, so that the position of equilibrium is unstable and physically impossible, and we are dealing with self-excitation.

Consider the special case where

$$e = F(i) = A + Bi + Ci^2 + Di^3 + Ei^4 + Fi^5.$$

We find then

$$R(a) = B + \tfrac{3}{4}Di^2 + \tfrac{5}{8}Fi^4.$$

We must assume that the coefficient $F > 0$, for otherwise beginning with a certain $a \geq a'$, R(a) will be negative and oscillations of amplitude $a_0 > a'$ will expand to infinity, which is ruled out physically.

Consider the equation

$$B + \tfrac{3}{4}Da^2 + \tfrac{5}{8}Fa^4 = 0,$$

whose solutions are

$$a_1^2 = -\frac{3}{5}\frac{D}{F} - \sqrt{\frac{9D^2}{25F^2} - \frac{8}{5}\frac{B}{F}}, \quad a_2^2 = \frac{-3}{5}\frac{D}{F} + \sqrt{}\,.$$

We will examine the three cases

$$(I)\ B>0,\ D>0: \quad (II)\ B>0,\ D<0: \quad (III)\ B<0.$$

In Case I, both roots are imaginary, R(a) is always positive, hence the system is dissipative. In Case II, if

$$B > \frac{9}{40}\frac{D^2}{F}$$

then the system is likewise dissipative. In the contrary case there may exist steady oscillations of amplitude a_2. However, the system is not self-exciting, and oscillations whose initial amplitudes are less than a_1 die down. In Case III, the system is self-exciting and there is a unique stationary régime for the oscillations with amplitude $a>a_2$.

15. We return to (6.3) and its approximate solution $x = a \sin\psi$ where a, ψ, are given by

$$\frac{da}{dt} = \Phi(a), \tag{15.1}$$

$$\frac{d\psi}{dt} = \omega(a), \tag{15.2}$$

where

$$\Phi(a) = -\frac{\varepsilon}{2\pi\nu}\int_0^{2\pi} f(a \sin\phi,\ a\nu\cos\phi)\cos\phi\, d\phi,$$

$$\omega(a) = \nu + \frac{\varepsilon}{2\pi\nu}a\int_0^{2\pi} f(\qquad\qquad\qquad)\sin\phi\, d\phi.$$

If we square ω and refer to (6.2) we obtain

$$(15.3)\quad \omega^2(a) = \frac{1}{\pi a}\int_0^{2\pi} F(a \sin \phi,\ a\nu \cos \phi)\sin \phi\, d\phi,$$

and likewise

$$(15.4)\quad \Phi(a) = -\frac{1}{2\pi\nu}\int_0^{2\pi} F(a \sin \phi,\ a\nu \cos \phi)\cos \phi\, d\phi.$$

Thus by means of (15.1), (15.2) the functions ω, Φ are determined directly in terms of the function F of (6.1).

16. We will now discuss (15.1) which determines the variation of the amplitude in function of the time. Observe that there must exist no $a^* > 0$ such that

$$\Phi(a) > 0 \text{ for } a > a^*,$$

For if such an a^* existed then taking an initial amplitude $a_0 > a^*$, we would obtain in view of (15.1) $a(t) \to +\infty$ $t \to \infty$ which is physically ruled out.

Referring to (15.1) we see that if the initial amplitude a_0 is not stationary, i. e., does not satisfy $\Phi(a) = 0$, then with increasing t the amplitude a (t) will steadily tend to a stationary determination.

The tendency of every oscillation to approach a steady oscillation points to the special role of steady oscillations for all high-frequency oscillatory processes. Indeed in such systems the intermediary régime tends very rapidly to a stationary régime and hence every oscillation may be viewed as practically stationary.

A noteworthy special case may be mentioned here where there are no intermediary régimes, and every oscillation is stationary. It will take place for example whenever the function F does not contain $\frac{dx}{dt}$ (conservative system). Then (6.1) may be written in the form

(16.1) $$\frac{d^2x}{dt^2} + F(x) = 0,$$

and so direct integration is possible. Indeed if we introduce the potential

$$U(x) = \int_0^x F(x)dx,$$

then (16.1) yields immediately

(16.2) $$\frac{1}{2}\left(\frac{dx}{dt}\right)^2 + U(x) = F = \text{const.}$$

Practically, however, this conservative case never occurs and there is always dissipation, hence loss of energy, or for that matter there may be self-oscillation and production of energy within the system.

17. We will now consider the stability of the stationary oscillations. Let a_1 be any root of $\Phi(a) = 0$. Then for a very near a_1 we will have $a = a_1 + \delta a$ and so from (15.1):

$$\frac{d\delta a}{dt} = \Phi'(a_1)\delta a.$$

This shows that a_1 is stable, that is to say, corresponds to a stable stationary oscillation if

(17.1) $$\Phi'(a_1) < 0,$$

while if

$$\Phi'(a_1) > 0,$$

then the corresponding stationary oscillation will be

unstable. In particular the static regime (a=0) will be unstable whenever

$$\Phi'(0) > 0, \tag{17.2}$$

and so this last inequality is the condition for self-excitation.

As we have already seen self-excitation is not necessary for the existence of self-oscillations in the system, that is for the existence of stable stationary oscillations. For that purpose there must merely exist an a_1 such that (17.1) holds.

An interesting case of frequent occurrence is where the system depends upon a parameter μ. An example is a series circuit with an impressed harmonic voltage of amplitude μ. Under the circumstances generally $\Phi(a)$ will be a function $\Phi(a, \mu)$. A typical situation is the graph $\Phi(a, \mu) = 0$ of Fig. 7. The dotted arcs represent the unstable stationary amplitudes, the heavy arcs the stable ones. Through the variation of μ there may thus arise "cyclic" régimes as indicated by the arrows.

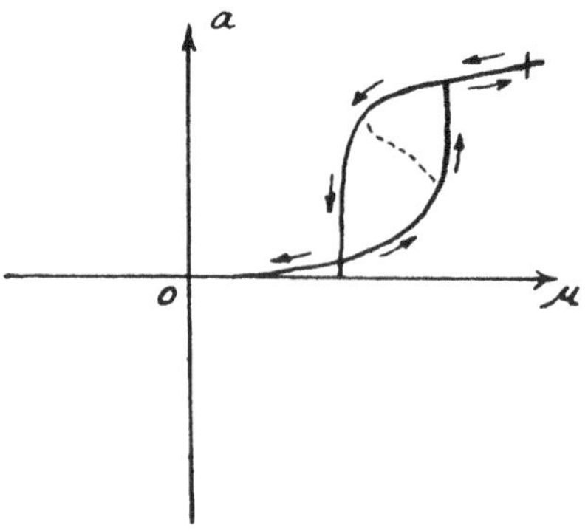

Fig. 7

18. Having discussed at length (15.1) and the amplitudes we will consider equation (15.2) for the frequency. Corresponding to a given frequency and hence to a given period, the frequency is $\frac{dx}{dt} = \omega(a)$, and generally frequency and period depend upon the amplitude. There are, however, important cases in practice when the oscillations do not depend upon the amplitude (isochronous system). An example is when we have identically

$$\int_0^{2\pi} f(a \sin \phi,\ a\nu \cos \phi) \sin \phi \, d\phi = 0.$$

This will occur notably if the initial equation is of one of two forms

$$\frac{d^2x}{dt^2} + \omega^2 x + \varepsilon f(x)\frac{dx}{dt} = 0, \tag{18.1}$$

$$\frac{d^2x}{dt^2} + \omega^2 x + \varepsilon F\left(\frac{dx}{dt}\right) = 0. \tag{18.2}$$

Notice that (18.2) can be reduced to the form (18.1) by the change of variable $\frac{dx}{dt} = y$.

Referring to (15) the solution of (18.1) will be in the first approximation

$$x = a \sin (\omega t + \phi), \tag{18.3}$$

where ϕ is constant and

$$\frac{da}{dt} = -\frac{a}{2}\lambda(a), \tag{18.4}$$

$$\lambda(a) = \frac{1}{\pi}\int_0^{2\pi} f(a \sin \tau) \cos^2\tau \, d\tau. \tag{18.5}$$

As an example equation (5.6) will have a solution described in the following relations

$$(18.6)\qquad \begin{cases} V = a \sin (t + \phi) \\ 2\sqrt{LC}\, \frac{da}{dt} = -\frac{L}{R}\, a + (M-DL)F(a), \\ F(a) = \frac{1}{\pi}\int_0^{2\pi} f(E_0 + a \sin \tau) \sin \tau \, d\tau. \end{cases}$$

III. REFINEMENT OF THE FIRST APPROXIMATION

19. In the present chapter we shall discuss a method for replacing the first approximation by one which will be somewhat more accurate, although, of coursé, essentially more complicated.

Let us examine again the exact equation (8.3) in a, ϕ. The process leading to the first approximation consisted essentially in replacing the right hand sides by their constant terms. One may think of these terms as corresponding to slow smooth variation and obtained by neglecting the rapidly changing terms represented by the trigonometric functions. In order to account to some extent for these more rapid changes, we will utilize the basic concept of the method of successive approximations.

Replace then first (8.3) by

$$\frac{d\bar{a}}{dt} = -\frac{\varepsilon}{\nu}\left\{K_o(a) + \sum(K_n(a)\cos n(\nu t+\phi) + L_n(a)\sin n(\nu t+\phi))\right\}$$

(19.1)

$$\frac{d\bar{\phi}}{dt} = \frac{\varepsilon}{\nu a}\left\{P_o(a) + \sum(P_n(a)\cos n(\nu t+\phi) + Q_n(a)\sin n(\nu t+\phi))\right\}$$

or which is the same by

$$\frac{d\bar{a}}{dt} = \frac{da}{dt} - \frac{\varepsilon}{\nu}\sum(K_n\cos n(\nu t+\phi) + L_n \sin n\ (\nu t+\phi)$$

(19.2)

$$\frac{d\bar{\phi}}{dt} = \frac{d\phi}{dt} + \frac{\varepsilon}{\nu}\sum(P_n\cos n(\nu t+\phi) + Q_n \sin n(\nu t+\phi).$$

In these relations a, ϕ are the solutions of the "smoothed out" equations (8.5), that is to say of the first approximations a, ϕ, in the sense that we have adopted so far.

Since a, ϕ, do not vary very rapidly, we will integrate the right hand sides of (19.2) as if the a, ϕ in the sums were constant. We thus obtain

$$\bar{a} = a - \frac{\varepsilon}{\nu}\sum\frac{K_n \sin n(\nu t+\phi) - L_n \cos n(\nu t+\phi)}{n\nu}$$

(19.3)

$$\bar{\phi} = \phi + \frac{\varepsilon}{\nu}\sum\frac{P_n \sin n(\nu t+\phi) - Q_n \cos n(\nu t+\phi)}{n\nu}.$$

The new refined first approximation will be given by

$$x = \bar{a}\sin(\nu t+\bar{\phi}).$$

After some simple calculations, this leads to

$$x = a \sin(\nu t+\phi) + \frac{\varepsilon}{\nu^2}\left\{ - f_o(a) + \sum_{n>1} \frac{f_n(a)\cos n(\nu t+\phi) + g_n(a)\sin n(\nu t+\phi)}{n^2-1}\right\} \tag{19.4}$$

where a, ϕ, are determined as before, and where f_n, g_n, are the Fourier coefficient in the expansion

$$f(a \sin \tau,\ a\nu \cos \tau) = f_o(a) + \sum (f_n(a) \cos n\tau + g_n(a) \sin n\tau). \tag{19.5}$$

We prove without difficulty

$$K_o = \tfrac{1}{2}f_1,\quad P_o = \tfrac{1}{2}g_1.$$

This enables us to put the relations for a, ϕ in the form

$$\frac{da}{dt} = -\frac{\varepsilon}{2\nu} f_1(a)$$

(19.6)

$$\frac{d\phi}{dt} = \omega(a) - \nu$$

where

$$\omega(a) = \nu + \frac{\varepsilon}{2\nu a} g_1(a) \tag{19.7}$$

or equivalently, to within quantities of the magnitude of ε^2,

$$\omega^2(a) = \nu^2 + \frac{\varepsilon}{a} g_1(a). \tag{19.8}$$

20. Since our results have been obtained by methods which have no pretension to rigor, it is necessary to examine directly the degree to which they satisfy our basic equation (6.3). We find immediately

$$(20.1)\qquad \frac{d^2x}{dt^2} + \nu^2 x = -\varepsilon \{f_o + \sum f_n \cos n(\nu t+\phi) + g_n \sin n(\nu t+\phi)\} + O(\varepsilon^2),$$

where $O(\varepsilon^2)$ denotes a quantity of the order of ε^2. By (19.6):

$$(20.2)\qquad \varepsilon f(x,\frac{dx}{dt}) = \varepsilon f(a \sin (\nu t+\phi), a\nu \cos (\nu t+\phi)) + O(\varepsilon^2).$$

Hence, finally

$$(20.3)\qquad \frac{d^2x}{dt^2} + \nu^2 x + \varepsilon f(x,\frac{dx}{dt}) = O(\varepsilon^2).$$

In order words the approximate solution (19.4) satisfies the initial equation (6.3) to within a quantity of the order of ε^2. More explicitly if $f(x, x')$ possesses partial derivations of order two or more and if in addition the amplitudes are bounded, then

$$\left|\frac{d^2x}{dt^2} + \nu^2 x + \varepsilon f(x,\frac{dx}{dt})\right| < K\varepsilon^2, \quad 0 \leq t < \infty,$$

where K is a constant which depends neither on ε nor on t.

Under these conditions then (19.4) satisfies (6.3) to within quantities of the magnitude of ε^2 and this uniformly in t for all non-negative t.

21. The question of the order of magnitude of the errors may also be treated in a different way. Namely we first make a change of variables in (6.3) and introduce new unknowns a, ψ, through the relations

$$x = a \sin \psi + \frac{\varepsilon}{\nu^2} \left\{ -f_0(a) + \sum_{n>1} \frac{f_n \cos n\psi + g_n \sin n\psi}{n^2 - 1} \right\}$$

(21.1)

$$\frac{dx}{dt} = a\omega(a) \cos \psi + \frac{\varepsilon}{2\nu} \left\{ -f_1(a) \sin \psi + \sum_{n>1} \frac{n\nu}{n^2-1} (g_n \cos n\psi - f_n \sin n\psi) \right\} .$$

After some simple computations it may be shown that these new variables satisfy the system

(21.2)
$$\frac{da}{dt} = -\frac{\varepsilon}{2\nu} f_1(a) + \varepsilon^2 X(a, \psi, \varepsilon)$$

$$\frac{d\psi}{dt} = \omega(a) + \varepsilon^2 Y(a, \psi, \varepsilon),$$

where X, Y are periodic functions of ψ (with period 2π), and regular with respect to ε in the neighborhood of $\varepsilon = 0$. Notice that (19.6) may be deduced from (21.2) by rejecting the terms in ε^2.

If we compare the refined approximation (19.4) with the earlier first approximation, we find that the latter merely represents the first harmonic in the Fourier series (19.4). The other harmonics will be of the order of magnitude of ε.

If we examine the stationary oscillations we find from (19.4) that they will be periodic with period $\frac{2\pi}{\omega(a)}$, where a is the corresponding stationary amplitude. The corresponding frequency is:

$$\omega(a) = \nu + \frac{\varepsilon}{2a\nu} g_1(a).$$

Since the relation between the frequency and the amplitudes is through the medium of a term proportional to ε, we may, if we continue to disregard terms in ε^2, replace in $\omega(a)$ the term in ε by any other which differs from it only by some term in ε^2; for instance, to within terms in ε^2 we may write

$$\omega(a) = \nu + \frac{\varepsilon}{2\nu x_{max}} g_1(x_{max}),$$

or also

$$\omega^2(a) = \nu^2 + \frac{\varepsilon}{x_{max}} g_1(x_{max}).$$

22. We will now apply the preceeding results to some examples.

(22.1) Example 1. Consider the differential equation for a conservative system:

$$\frac{d^2x}{dt^2} + \nu^2 x + \varepsilon f(x) = 0, \tag{22.2}$$

where f(x) is an odd function. We then verify that in (19.5) only the g_n terms remain. As a consequence the approximate solution (19.4) assumes the form

(22.3) $$x = a \sin(\omega t+\theta) + \frac{\varepsilon}{\nu^2} \sum_{n>1} \frac{g_n(a) \sin n(\omega t+\theta)}{n^2-1}$$

(22.4) $$\omega = \nu + \frac{\varepsilon}{2\nu a} g_1(a),$$

(22.5) $$\omega^2 = \nu^2 + \frac{\varepsilon}{a} g_1(a),$$

where a, θ are arbitrary constants.

As a special case suppose that we are dealing with

$$\frac{d^2x}{dt^2} + \nu^2 x + \varepsilon x^3 = 0.$$

Since $f(x) = x^3$, we find

$$f(a \sin \tau) = \frac{3}{4}a^2 \sin \tau - \frac{1}{4}a^3 \sin 3\tau,$$

and hence the refined first approximation will be

$$x = a \sin(\omega t+\theta) - \frac{\varepsilon a^3}{32} \sin(3\omega t+\theta),$$

where

$$\omega = \nu + \frac{3}{8}\varepsilon a^2.$$

(22.6) Example 2. Let the basic equation be the approximate equation (11.1) for the pendulum without friction. To reduce this relation to the form (22.2) we set

(22.7) $$\frac{g}{l} = \nu^2, \frac{-g}{6l}x^3 = \varepsilon f(x)$$

and (22.3), (22.4) yield here

$$x = a \sin (\omega t+\theta) + \frac{a^3}{192} \sin 3(\omega t+\theta),$$

(22.8)

$$\omega = \sqrt{\frac{g}{l}} \left(1 - \frac{a^2}{16}\right).$$

The comparison with the classical series for the same quantities shows that we are just obtaining the first terms of their series.

(22.9) Example 3. As our next example, we will take equation (2.2) for the oscillating shaft. We choose $c(\theta)$ as in (11.5), and find as our basic solution

$$x = a \sin (\omega t+\theta) + \frac{4h}{\pi k} \sum_{n_{odd}} \frac{\sin n(\omega t+\theta)}{n(n^2-1)}$$

(22.10)

$$\omega = \sqrt{\frac{J_1 + J_2}{J_1 J_2}}\, k\left(1 + \frac{2h}{\pi ka}\right) = \nu\left(1 + \frac{2h}{\pi ka}\right).$$

In the present case as it happens, it is not difficult to obtain the exact solution, and it is found to be

(22.11) $$x = a \sin (\omega t+\theta) + \frac{4h}{\pi k} \sum_{n_{odd}} \frac{\sin n(\omega t+\theta)}{n\left[\left(\frac{\omega}{\nu}\right)^2 n^2-1\right]},$$

(22.12)

$$\omega = \nu\sqrt{1 + \frac{4h}{\pi ka}}.$$

In this case then the approximate solution may be obtained by replacing in the denominators

$$n\left[\left(\frac{w}{\nu}\right)^2 n^2 - 1\right]$$

the frequency ω by its approximation ν. Moreover we see that (22.11) yields the accurate expression of the first two terms of the exact solution considered as a power series in $\frac{h}{ka}$.

23. We will now consider the approximate solution (19.4) as applied to a dissipative oscillatory system

$$\frac{d^2x}{dt^2} + \nu^2 x + \epsilon f(x) \frac{dx}{dt} = 0. \tag{23.1}$$

We have here

$$f(x, \frac{dx}{dt}) = f(x)\frac{dx}{dt} . \tag{23.2}$$

Therefore

$$f(a \sin \tau,\ a\nu \cos \tau) = f\ (a \sin \tau)\ a\nu \cos \tau. \tag{23.3}$$

Introduce the function

$$F(x) = \int_0^x f(x)dx, \tag{23.4}$$

and form the Fourier series:

$$F(a \cos \phi) = \sum F_n^*(a) \cos n\phi \tag{23.5}$$

By differentiating both sides of (23.5) we obtain in combination with (23.4):

$$af(a \cos \phi) \sin \phi = \sum nF_n^*(a) \sin n\phi.$$

If we set $\phi = \tau + \frac{3\pi}{2}$, (23.3) yields:

$$f(a \sin \tau,\ a\nu \cos \tau) = -\nu \sum nF_n^* \sin n(\nu t + \phi + \frac{3\pi}{2}) .$$

Hence (19.4) yields the approximate solution

$$(23.6)\quad x = a \sin (\nu t+\phi) - \frac{\varepsilon}{\nu}\sum_{n>1} \frac{n}{n^2-1} F_n^*(a)\sin n(\nu t+\phi+\tfrac{3\pi}{2}),$$

where ϕ is an arbitrary phase constant. Here a satisfies

$$\frac{da}{dt} = - \frac{\varepsilon}{2} F_1^*(a).$$

In particular the stationary amplitudes are the roots of

$$F_1^*(a) = 0.$$

If instead of ϕ we introduce another constant phase $\theta = \phi - \frac{\pi}{2}$, then (23.6) takes the form

$$(23.7)\quad x = a \cos (\omega t+\theta) - \frac{\varepsilon}{\nu}\sum_{n>1} F_n^*(a)\sin n(\omega t+\theta)$$

where $\omega = \nu$.

Let us apply the argument to van der Pol's equation (4.4). We have then

$$f(x) = x^2 - 1,\ F(x) = \frac{x^3}{3} - x,$$

and hence

$$F(a \cos \phi) = \frac{a^3\cos^3\phi}{3} - a \cos \phi = a(\frac{a^2}{4} - 1)\cos \phi + \frac{a^3}{12}$$

$$\cos 3\phi,$$

so that

$$(23.8)\quad \begin{cases} F_1^*(a) = a(\frac{a^2}{4} - 1),\ F_3^*(a) = \frac{a^3}{12}, \\ F_n^*(a) = 0 \text{ for } n \neq 1,\ 3. \end{cases}$$

Thus (23.7) becomes here

$$(23.9) \qquad x = a\cos(t+\theta) - \frac{\varepsilon a^3}{32}\sin 3(t+\theta)$$

where θ is an arbitrary constant and a satisfies

$$\frac{da}{dt} = \frac{\varepsilon a}{2}\left(1 - \frac{a^2}{4}\right).$$

For the stationary oscillations a = 2, and hence

$$(23.10) \qquad x = 2\cos(t+\theta) - \frac{\varepsilon}{4}\sin 3(t+\theta).$$

24. Returning to (23.1) we notice that in the approximation under consideration, the frequency is ν, that is to say the first term in the expansion of the frequency in powers of ε. The second term, the term in ε, is 0 in this case. We will now show how to calculate for stationary oscillations the term in ε^2 by means of (23.7).

We first observe that since a stationary oscillation is periodic with a certain period T, we may expand the exact solution x in a Fourier series:

$$(24.1) \quad x = a\cos(\omega t+\theta) + \sum_{n>1} A_n \cos n(\omega t+\theta) + B_n \sin n(\omega t+\theta),$$

where a is the amplitude, θ the phase of the first harmonic and $\omega = \frac{2\pi}{T}$. On the other hand (23.1) yields:

$$\int_0^T \left(\frac{d^2x}{dt^2}x + \nu^2x^2 + \varepsilon f(x)x\frac{dx}{dt}\right)dt = 0.$$

Since x is periodic we have identically

$$\int_0^T \frac{d^2x}{dt^2}x\,dt = -\int_0^T \left(\frac{dx}{dt}\right)^2 dt, \quad \int_0^T f(x)x\,dx = 0.$$

Hence

$$\int_0^T \left(\frac{dx}{dt}\right)^2 dt = \nu^2 \int_0^T x^2 dt. \tag{24.2}$$

Substituting (24.1) into (24.2) we obtain

$$\omega^2\left(a^2 + \sum_{n>1} n^2(A_n^2 + B_n^2)\right) = \nu^2\left(a^2 + \sum_{n>1} (A_n^2 + B_n^2)\right),$$

and hence

$$\left(\frac{\omega}{\nu}\right)^2 = \frac{a^2 + \sum_{n>1} (A_n^2 + B_n^2)}{a^2 + \sum_{n>1} n^2(A_n^2 + B_n^2)} \tag{24.3}$$

By comparison of (23.7) with (24.1) we see that approximately

$$A_n = 0, \quad B_n = \frac{-\varepsilon n}{\nu(n^2-1)} F_n^*(a).$$

We have, therefore, the following expression for the frequency of the stationary oscillations:

$$\left(\frac{\omega}{\nu}\right)^2 = \frac{1 + \frac{\varepsilon^2}{\nu^2 a^2} \sum_{n>1} \left(\frac{n}{n^2-1} F_n^*(a)\right)^2}{1 + \frac{\varepsilon^2}{\nu^2 a^2} \sum_{n>1} n^2\left(\frac{n}{n^2-1} F_n^*(a)\right)^2}. \tag{24.4}$$

It is not difficult to see that (24.4) will hold in any case to within terms of order ε^3. Neglecting therefore in (24.4) terms in ε^4 we obtain the simpler expression

$$(24.5) \qquad \left(\frac{\omega}{\nu}\right)^2 = 1 - \frac{\varepsilon^2}{\nu^2}\sum_{n>1}\frac{n^2}{n^2-1}\left(\frac{F_n^*(a)}{a}\right)^2 ,$$

or finally:

$$(24.6) \qquad \omega = \nu - \frac{\varepsilon}{2\nu}\sum_{n>1}\frac{n^2}{n^2-1}\left(\frac{F_n^*(a)}{a}\right)^2 .$$

Let us apply this formula to van der Pol's equation (4.4). The stationary oscillations correspond to a = 2 and so from (23.8) there follows:

$$F_3^*(a) = \frac{a^3}{12} = \frac{2}{3}; \; F_n^*(a) = 0, \; n \neq 3.$$

Hence

$$(24.7) \qquad \omega = 1 - \frac{\varepsilon^2}{16} .$$

IV. CONSTRUCTION OF THE HIGHER APPROXIMATIONS

25. In the preceding chapter we indicated a way to improve the first approximation by taking into consideration the higher harmonics, and we have shown that the basic equation (6.3) was satisfied to within terms in ε^2. Moreover for equations of the form (23.1) we gave a method for computing approximately the stationary frequencies to within terms in ε^3. We shall now consider methods for forming approximate solutions corresponding to stationary oscillations which satisfy (6.3) to within

terms in any given power of ε. Once and for all we will assume that the functions entering in (6.3) have all the required derivatives and that the amplitudes under consideration are bounded.

Whenever we state that an expression satisfies the differential equation to within terms of ε^m, we will always understand thereby that the error is of order ε^m uniformly in t for all non-negative t.

26. Consider first the conservative system

$$\frac{d^2x}{dt^2} + \nu^2 x + \varepsilon f(x) = 0. \tag{26.1}$$

In this case clearly an "arbitrary" oscillation will be stationary. Referring to (22.3) the refined first approximation will be

$$x = a \sin(\omega t+\theta) + \frac{\varepsilon}{\nu^2} h_o(a) \tag{26.2}$$

$$+ \sum_{n>1} \frac{g_n \sin n(\omega t+\theta) + h_n \cos n(\omega t+\theta)}{n^2 - 1}$$

where a, θ are arbitrary constants, and h_n, g_n are the Fourier coefficients in the expansion

$$f(a \sin \tau) = \sum (h_n \sin n\tau + g_n \cos n\tau). \tag{26.3}$$

Moreover here

$$\omega^2 = \nu^2 + \frac{\varepsilon}{a} g_1(a). \tag{26.4}$$

We will modify (26.2) so as to express it in terms of the coefficients of the Fourier expansion

$$(26.5) \qquad f(a \cos \tau) = \sum f_n(a) \cos n\tau.$$

For this purpose replace in (26.3) τ by $\tau + \frac{\pi}{2}$, thus obtaining

$$f(a \cos \tau) = \sum (g_n \sin n(\tau + \tfrac{\pi}{2}) + h_n \cos n(\tau + \tfrac{\pi}{2})).$$

The identification with (26.5) yields

$$(26.6) \qquad g_n = f_n \sin \frac{n\pi}{2}, \; h_n = f_n \cos \frac{n\pi}{2} .$$

Substituting these expressions in (26.2) we find:

$$x = a \sin (\omega t + \theta) + \frac{\varepsilon}{\nu^2} \left\{ -f_o + \sum_{n>1} \frac{f_n \cos n\,(\omega t + \theta - \frac{\pi}{2})}{n^2 - 1} \right\}$$

and hence replacing θ by the new arbitrary $\phi = \theta - \frac{\pi}{2}$, we have:

$$(26.7) \quad x = a \cos (\omega t + \phi) + \frac{\varepsilon}{\nu^2} \left\{ -f_o + \sum_{n>1} \frac{f_n \cos n(\omega t + \phi)}{n^2 - 1} \right\} .$$

This expression is more convenient than (26.2) in that it contains only cosine terms.

In view of (26.6), formula (26.4) becomes here

$$(26.8) \qquad \omega^2 = \nu^2 + \varepsilon \frac{f_1(a)}{a} .$$

27. The expressions (26.7), (26.8) suggest the following method for obtaining approximations of any order. Represent the solution of (26.1) in the form $x = z(\tau)$, where $\tau = \omega t + \phi$, with ϕ an arbitrary constant and $z(\tau)$ a periodic function of τ with period 2π. Notice that $x = z(\tau)$ will satisfy (26.1) if, and only if $z(\tau)$ satisfies the equation

$$\omega^2 \frac{d^2 z}{d\tau^2} + \nu^2 z + \varepsilon f(z) = 0. \tag{27.1}$$

We will endeavor to obtain a solution of (27.1) such that we have expansions

$$\begin{aligned} z(\tau) &= z_0(\tau) + \varepsilon z_1(\tau) + \dots \\ \omega^2 &= \alpha_0 + \alpha\varepsilon_1 + \dots \end{aligned} \tag{27.2}$$

where the coefficients are to be determined by substituting in (27.1) and annulling the powers of ε. Furthermore this is to be done in such a way that the z_n are periodic in τ with period 2π.

We thus obtain the following recursive relations:

$$\left\{\begin{aligned} &\alpha_0 \frac{d^2 z_0}{d\tau^2} + \nu^2 z_0 = 0, \\ &\alpha_1 \frac{d^2 z_1}{d\tau^2} + \nu^2 z_1 = -f(z_0) - \alpha_1 \frac{d^2 z_0}{d\tau^2} \\ &\dots\dots\dots\dots\dots\dots \\ &\alpha_0 \frac{d^2 z_{n+1}}{d\tau^2} + \nu^2 z_{n+1} = F(z_0, \dots, z_n) \\ &\qquad -\alpha_{n+1} \frac{d^2 z_0}{d\tau^2} - \dots - \alpha_1 \frac{d^2 z_n}{d\tau^2} \end{aligned}\right. \tag{27.3}$$

where $F(z_0, z_1, \dots, z_n)$ is a polynomial in

$z_1, \dots z_n$.

Suppose that $z_0, z_1, \dots, z_k$ satisfy the first $k + 1$ relations of the system. Then clearly

$$(27.4) \qquad x = z_0 + \varepsilon z_1 + \dots + \varepsilon^k z_k$$

where

$$(27.5) \qquad \omega^2 = \alpha_0 + \varepsilon\alpha_1 + \dots + \varepsilon^k \alpha_k,$$

will satisfy our initial equation (26.1) to within terms of order ε^{k+1}. Thus x may be considered as the required approximation to this order.

The successive determinations of the coefficients z_n, α_n contain arbitrary elements which we will utilize to remove the secular terms in the solution. Take first,

$$(27.6) \qquad z_0 = a \cos \tau, \ \alpha_0 = \nu^2,$$

as the solution of the first equation (27.3). The second equation yields then

$$(27.7) \quad \nu^2\left(\frac{d^2 z_1}{d\tau^2} + z_1\right) = -f(a \cos \tau) + \alpha_1 a \cos \tau.$$

Hence in view of (26.5) we have:

$$(27.8) \quad \nu^2\left(\frac{d^2 z_1}{d\tau^2}+z_1\right) = -f_0(a) - \sum_{n\geq 2} f_n \cos n\tau+(\alpha_1 a-f_1)\cos \tau.$$

To avoid secular terms there must be no terms in $\cos \tau$ at the right and so we must have

$$(27.9) \qquad \alpha_1 = \frac{f_1(a)}{a}.$$

From this follows for the solution of (27.8):

$$(27.10) \qquad z_1 = -\frac{f_o}{\nu^2} + \frac{1}{\nu^2}\sum_{n>1} \frac{f_n \cos n\tau}{n^2-1} .$$

In the same way and by an evident induction one may obtain every z_n and avoid step by step the presence of secular terms; the details may be left to the reader.

28. As an application take the equation

$$(28.1) \qquad \frac{d^2x}{d\tau^2} + x + \varepsilon x^3 = 0.$$

We find here

$$(28.2) \qquad \left\{ \begin{array}{l} \dfrac{d^2z_1}{d\tau^2} + z_1 = -z_o^3 - \alpha_1 \dfrac{d^2z_o}{d\tau_o^2} , \\ - - - - - - - - - - - - - - - \end{array} \right.$$

$$(28.3) \qquad z_o = a \cos \tau, \ \alpha_1 = 1.$$

In view of (28.3) the first relation (28.2) becomes

$$\frac{d^2z_1}{d\tau^2} + z_1 = (\alpha_1 a - \tfrac{3}{4} a^3) \cos \tau - \frac{a^3}{4} \cos 3\tau.$$

Therefore

$$(28.4) \qquad \alpha_1 = \tfrac{3}{4}a^2, \ z_1 = \frac{a^3}{32} \cos 3\tau.$$

From this follows by the regular application of the method:

$$(28.5)\quad \alpha_2 = \frac{3a^4}{128},\quad z_2 = \frac{-21}{1024}a^5\cos 3\tau + \frac{a^5}{1024}\cos 5\tau,$$

and finally to within terms in ε^3

$$(28.6)\quad x = a\cos(\omega t+\phi) + \frac{\varepsilon a^3}{32}\left(1 - \varepsilon\frac{21}{32}\right)\cos 3(\omega t+\phi)$$

$$+ \varepsilon^2 \frac{a^5}{1024}\cos 5(\omega t+\phi),$$

where a, ϕ are arbitrary constants and ω is given by

$$(28.7)\qquad \omega^2 = 1 + \tfrac{3}{4}\varepsilon a^2 + \frac{3}{128}\varepsilon^2 a^4.$$

The same method may be applied in an obvious way to

$$(28.8)\quad \frac{d^2x}{dt^2} + \nu^2 x + \varepsilon f(x) + \varepsilon^2 f_1(x) + \ldots = 0.$$

29. Consider again the system (26.1) with f(x) a power series in x:

$$(29.1)\qquad f(x) = b_2x^2 + b_3x^3 + \ldots\ldots$$

Here there is no small parameter ε. However if we merely wish to consider small oscillations then clearly f(x) will be small with respect to ν^2 and furthermore it consists of a series of terms of increasing orders of small magnitude. This justifies to a certain extent the following procedure. Replace (26.1) by

$$(29.2)\quad \frac{d^2x}{dt^2} + \nu^2 x + \rho b_2 x^2 + \rho^2 b_3 x^3 + \ldots = 0$$

where we will consider ρ as a small parameter. This equation is now solved as before as a power series in ρ, after which the parameter ρ is made equal to unity, thus yielding an approximate solution.

We will now consider analogous methods for a general non-conservative system (6.3). We first modify our relations (19.4), (19.5), (19.6) for the refined first approximation by setting $\nu t+\phi = \frac{\pi}{2}+\psi$ and thus obtaining:

$$(29.3)\quad x = a\cos\psi + \frac{\varepsilon}{\nu^2}\Big\{-F_0(a)+\sum_{n>1}\frac{F_n\cos n\psi+G_n\sin n\psi}{n^2-1}\Big\}$$

$$(29.4)\qquad \frac{da}{dt} = \frac{\varepsilon}{2\nu}G_1(a),\quad \frac{d\psi}{dt} = \omega(a),$$

$$(29.5)\qquad \omega(a) = \nu + \frac{\varepsilon}{2\nu a}F_1(a),$$

where F_n, G_n are the Fourier coefficients in the expansion:

$$(29.6)\quad f(a\cos\tau,\ -a\nu\sin\tau) = \sum(F_n(a)\cos n\tau+G_n(a)\sin n\tau).$$

For the stationary oscillations we have

$$(29.7)\qquad G_1(a) = 0,$$

$$(29.8)\qquad \psi = \omega(a)t+\phi,$$

where ϕ is an arbitrary constant. Thus (29.3) may be written more explicitly as

$$(29.9)\qquad x = a\cos(\omega(a)t + \phi) + \frac{\varepsilon}{\nu^2}\Big\{-F_0(a) + \sum_{n>1}\frac{F_n\cos n(\omega(a)t + \phi) + G_n\sin n(\omega(a)t + \phi)}{n^2 - 1}\Big\}.$$

For conservative systems as we have seen $G_1(a)$ is identically 0 and hence the approximate solution (29.9) contains the two arbitrary constants a, ϕ. We will now consider a case where $G_1(a)$ is not identically 0 in any interval of the variable a. Assume that $G_1(a)$ has only simple roots, so that if for a certain a: $G_1(a) = 0$, then the corresponding $G'(a) \neq 0$. Referring to (29.7), (29.9) we see that to every root of $G_1(a)$ there corresponds a certain stationary régime, and that for this régime the expression (29.7) depends upon the single arbitrary constant ϕ.

30. We will now take up the higher approximations for non-conservative systems and in the main use the same methods as for conservative systems. Write down the solution of (6.3) corresponding to stationary oscillations in the form

$$(30.1)\qquad x = z(\omega t + \phi)$$

where ϕ is an arbitrary constant, ω the frequency, and $z(\tau)$ a periodic function with the period 2π.

We first observe that $z(\tau)$ must satisfy the differential equation

$$(30.2)\qquad \omega^2\frac{d^2z}{d\tau^2} + \nu^2 z + \varepsilon f\left(z, \omega\frac{dz}{d\tau}\right) = 0.$$

We now endeavor to obtain $z(\tau)$ and ω as power series in ε:

$$(30.3)\qquad \begin{cases} z(\tau) = z_0(\tau) + \varepsilon z_1(\tau) + \cdots \\ \omega = \omega_0 + \varepsilon\omega_1 + \cdots, \end{cases}$$

where z_n is a periodic function with the period 2π. Proceeding as before by substitution in (30.2) we obtain a recursive system

$$(30.4)\qquad \begin{cases} \omega_0^2 \dfrac{d^2 z_0}{d\tau^2} + \nu^2 z_0 = 0 \\ \omega_0^2 \dfrac{d^2 z_1}{d\tau^2} + \nu^2 z_1 = -f(z_0\omega_0 \dfrac{dz_0}{d\tau} - 2\omega_0\omega_1 \dfrac{d^2 z_0}{d\tau^2} \\ \cdots\cdots\cdots\cdots\cdots \end{cases}$$

where at the right in the nth equation there are only terms in $z_0, \ldots, z_{n-1}$ and their derivatives as well as in $\omega_0, \ldots, \omega_{n-1}$. The first equation is solved as

$$(30.5)\qquad z_0 = a \cos \tau, \quad \omega_0 = \nu,$$

where a is as yet an indeterminate constant. Substituting in the second relation (30.4) there comes:

$$(30.6)\qquad \nu^2(\frac{d^2 z_1}{d\tau^2} + z_1) = -\sum (F_n \cos n\tau + G_n \sin n\tau) + 2\nu\omega_1 a \cos \tau.$$

To avoid secular terms we must have

$$(30.7)\qquad G_1(a) = 0, \quad \omega_1 = \frac{F_1(a)}{2\nu a},$$

which determine a and ω_1. This solves (30.6) as

$$z_1 = a_1 \cos \tau + \frac{1}{\nu^2} \left\{ -F_o(a) + \sum_{n>1} \frac{F_n \cos n\tau + G_n \sin n\tau}{n^2 - 1} \right\}, \tag{30.8}$$

where a_1 is an indeterminate constant. Notice in particular that in contrast to conservative systems, z_1 is not fully determined at the first step. For the amplitude a_1 of the first harmonic will be determined by the condition that z_2 be free from secular terms.

As a consequence of $G_1'(a) \neq 0$ and the other assumption made regarding $G_1(a)$ the process may be continued indefinitely.

31. It is to be observed that in the method just exposed a_n and ω_{n+1} are determined at the same step. In other words, ω_{n+1} is determined at the same time as the function $z_n(\tau)$. For this reason the joint determinations should be for

$$\left\{ \begin{aligned} x &= z_o(\omega t+\phi) + \ldots + \varepsilon^N z_N(\omega t+\phi), \\ \omega &= \nu + \ldots\ldots + \varepsilon^{N+1}\omega_{N+1} \end{aligned} \right. \tag{31.1}$$

that is to say with x up to the order ε^N and ω to the order ε^{N+1}. For instance if N = 0 then

$$x = a \cos (\omega t + \phi),$$

$$\omega = \nu + \frac{\varepsilon}{2\nu a} F_1(a), \quad G_1(a) = 0$$

which is our first approximation, obtained previously by the averaging principle. For N = 1 we have

$$x = (a+\varepsilon a_1)\cos(\omega t+\phi) + \frac{\varepsilon}{\nu^2}\Big\{-F_0(a) + \sum_{n>1}\frac{F_n(a)\cos n(\omega t+\phi) + G_n(a)\sin n(\omega t+\phi)}{n^2-1}\Big\} \tag{31.2}$$

$$\omega = \nu + \frac{\varepsilon}{2\nu}F_1(a) + \varepsilon^2\omega_2 .$$

By and large, the results are the same as before as far as the first step of approximation goes, except that they are now obtained systematically and not by some special device.

32. We will now consider the same questions not merely for stationary oscillations but also for the general case, that is to say, for oscillations which need not be stationary.

The form of the refined first approximation suggests looking for a solution of (6.3) of the form

$$x = z(\psi, a), \tag{32.1}$$

where $z(\psi,a)$ is a periodic function of ψ with period 2π, and where

$$\frac{da}{dt} = A(a),\ \frac{d\psi}{dt} = \omega(a). \tag{32.2}$$

By differentiating and substituting in (6.3) we obtain

$$\frac{\partial^2 z}{\partial\psi^2}\omega^2 + 2\frac{\partial^2 z}{\partial\psi\partial a}\omega A + \frac{\partial^2 z}{\partial a^2}A^2 + \frac{\partial z}{\partial\psi}\frac{\partial\omega}{\partial a}A + \frac{\partial z}{\partial a}\frac{\partial A}{\partial a}A + \nu^2 z + \varepsilon f\Big(z, \frac{\partial z}{\partial\psi}\omega + \frac{\partial z}{\partial a}A\Big) . \tag{32.3}$$

It is clear that if we find z, A, ω, satisfying (32.3) to within any particular order of magnitude in ε then (32.1), provided that (32.2) holds, will satisfy (6.3) to within the same order. To find the required expressions of z, A, ω, we will set

$$(32.4)\quad \begin{cases} z(\psi, a) = z_0(\psi, a) + \varepsilon z_1(\psi, a) + \dots \\ A(a) = \varepsilon A_1(a) + \varepsilon^2 A_2(a) + \dots \\ \omega(a) = \nu + \varepsilon \Omega_1(a) + \dots \end{cases}$$

where we assume $z_n(\psi, a)$ periodic in ψ with the period 2π. These expressions are substituted in (32.3) and yield the system:

$$(32.5)\quad \begin{cases} \dfrac{\partial^2 z_0}{\partial \psi^2} + z_0 = 0 \\ \left(\dfrac{\partial^2 z_1}{\partial \psi^2} + z_1\right)\nu^2 = -f\left(z_0, \dfrac{\partial z_0}{\partial \psi}\right) - 2\nu\Omega_1 \dfrac{\partial^2 z_0}{\partial \psi^2} \\ \qquad\qquad -2\nu A_1 \dfrac{\partial^2 z_0}{\partial \psi \partial a} \end{cases}$$

. . . .

The first is solved as

$$(32.6)\qquad z_0 = a \cos \psi.$$

We could equally start with any other solution, for instance $z_0 = a \sin \psi$, but this would not introduce any essential change anywhere. Substituting then in the second equation of (32.5) we obtain:

$$(32.7)\quad \left(\frac{\partial^2 z_1}{\partial \psi^2} + z_1\right)\nu^2 = 2\nu(\Omega_1 a \cos \psi + A_1 \sin \psi) - \sum (F_n \cos n\psi + G_n \sin n\psi).$$

To avoid secular terms, we must have

$$(32.8) \qquad 2\nu\Omega_1 a_1 = F_1(a), \quad 2\nu A_1 = G_n(a)$$

and the resulting equation for z_1 is solved as:

$$(32.9) \quad z_1 = \frac{1}{\nu^2}\left\{-F_o(a) + \sum_{n>1} \frac{F_n(a)\cos n\psi + G_n(a)\sin n\psi}{n^2 - 1}\right\}.$$

The process continues in the obvious way. We thus obtain in succession

$$z_o, z_1, \ldots, z_n; A_1, \ldots, A_n; \Omega_1 \ldots, \Omega_n$$

up to any index n. For instance if we have reached the value $n = N$, then we obtain a solution to within the order ε^{N+1} of the form

$$(32.10) \quad x = a\cos\psi + \varepsilon z_1(\psi, a) + \ldots + \varepsilon^N z_N(\psi, a),$$

where a, ψ, satisfy

$$(32.11) \quad \begin{aligned} \frac{da}{dt} &= \frac{\varepsilon}{2\nu}G_1(a) + \varepsilon^2 A_2(a) + \ldots + \varepsilon^N A_N(a), \\ \frac{d\psi}{dt} &= \nu + \frac{\varepsilon}{2\nu a}F_1(a) + \varepsilon^2\Omega_2(a) + \ldots + \varepsilon^N\Omega_N(a). \end{aligned}$$

Thus for $N = 1$ we obtain precisely the formulas for the refined first approximation.

33. Referring now to the first equation (32.11) we see that the stationary amplitudes are given to within order ε^{N+1} by the solutions of the equation:

$$(33.1) \qquad \frac{\varepsilon}{2\nu}G_1(a) + \ldots + \varepsilon^N A_N(a) = 0.$$

Let a_0 be a root of $G_1(a)$. Under our assumptions it is not a double root, and so (33.1) may be solved in the form

$$a = a_0 + a_1\varepsilon + \dots .$$

In the first approximation we will have

$$a_1 = \frac{-2\nu A_2(a_0)}{G_1'(a_0)} .$$

The stationary régime under consideration will be stable if

$$\frac{\varepsilon}{2\nu}G_1'(a) + \varepsilon^2 A_2'(a) + \dots + \varepsilon^N A_N'(a) < 0,$$

and unstable otherwise. Since clearly the left hand side is

$$\frac{\varepsilon}{2\nu}G_1'(a_0) + O(\varepsilon^2)$$

the question of stability for ε small enough will depend upon the sign of $G_1'(a_0)$, that is to say we obtain the same criterion as for the first approximation.

Observe also that the equations for the Nth approximation like those of the first, show that the amplitude a will increase or decrease monotonely approaching from above or from below the nearest stationary amplitude according to the sign of $\frac{da}{dt}$ for $t = 0$.

In general, one must emphasize the fact that, except for certain singular cases, the relations for the first approximation provide the same qualitative indications for the starting of self-oscillations as the higher approximations. Generally speaking, the higher approximations provide quantitative rather than new qualitative information. In view of this and of the difficulty of computing the higher approximation, it is usually quite sufficient to obtain the first approximation.

V. LINEARIZATION

34. In the present chapter, we will first of all endeavor to obtain suitable interpretations for the equations of the first approximation. We begin by writing the basic differential equation in the form

$$m\frac{d^2x}{dt^2} + kx + \varepsilon f(x, \frac{dx}{dt}) = 0, \tag{34.1}$$

where m, k, are positive. This system has two well known interpretations, the one mechanical, the other electrical.

We have obtained as the first approximation a solution

$$x = a \cos \psi, \tag{34.2}$$

where a, ψ satisfy

$$\begin{cases} \dfrac{da}{dt} = \dfrac{\varepsilon}{2\pi \nu m} \displaystyle\int_0^{2\pi} f(a \cos \phi, -a\nu \sin \phi) \sin \phi \, d\phi \\ \dfrac{d\psi}{dt} = \omega(a) \end{cases} \tag{34.3}$$

$$(34.4)\quad \nu^2 = \frac{k}{m},\ \omega^2(a) = \nu^2 + \frac{\varepsilon}{\pi m a}\int_0^{2\pi} f(a\cos\phi, -a\nu\sin\phi)\cos\phi\, d\phi.$$

It is to be kept in mind also that the first approximation (34.2) represents the fundamental harmonic in the expression of the refined first approximation (see for instance formula (29.3)) which satisfy (34.1) to within order ε^2.

Let us introduce the functions of the amplitude $\bar{k}(a)$, $\bar{\lambda}(a)$ defined by

$$(34.5)\quad \bar{\lambda} = \frac{-\varepsilon}{\pi a\nu}\int_0^{2\pi} f(a\cos\phi, -a\nu\sin\phi)\sin\phi\, d\phi$$

$$(34.6)\quad \bar{k} = k + \frac{\varepsilon}{\pi a}\int_0^{2\pi} f(a\cos\phi, -a\nu\sin\phi)\cos\phi\, d\phi.$$

In terms of these quantities the equations (34.3) for the first approximation take the form

$$(34.7)\quad \frac{da}{dt} = -\frac{\bar{\lambda}}{m}a,\qquad \frac{d\psi}{dt} = \omega = \sqrt{\frac{\bar{k}}{m}}\,.$$

As a consequence, we obtain by a direct if lengthy computation:

$$m\frac{d^2x}{dt^2} + \bar{\lambda}\frac{dx}{dt} + \bar{k}x = O(\varepsilon^2)\,.$$

We may then say that the first approximation (34.2) under consideration satisfies to within the order ε^2 the linear equation

(34.8) $$m \frac{d^2x}{dt^2} + \bar{\lambda} \frac{dx}{dt} + \bar{k}x = 0.$$

In short in the first approximation the oscillations of the non-linear system under consideration are equivalent to those of a linear system with a dissipation coefficient $\bar{\lambda}$ and a spring constant $\bar{k}$. The approximation is to the order ε^2, that is to say neglecting quantities of the same order as when we formed the first approximation. For this reason we will call $\bar{\lambda}$ the equivalent dissipation coefficient, and $\bar{k}$ the equivalent spring constant. The linear system (34.8) will also be said to be equivalent to the assigned system. From the comparison of (34.8) with the given equation (34.1) we see that the former arises from the non-linear system by replacing the non-linear term or restoring force of the mechanical analogy

(34.9) $$F = \varepsilon f(x, \frac{dx}{dt})$$

by the linear term

(34.10) $$F_1 = k_1 x + \bar{\lambda} \frac{dx}{dt} ,$$

where $k_1 = \bar{k} - k$.

Let us remark also that $\bar{\delta} = \frac{\bar{\lambda}}{2m}$ is the dissipation-decrement in the equivalent linear circuit, and $\omega = \sqrt{\frac{\bar{k}}{2m}}$ the proper period of its oscillations, to within the order of ε^2.

35. We may conclude then that the equations (34.7) of the first approximation may be derived as follows: Linearize the system by substituting for the restoring F of (34.9) the restoring force F_1 of (34.10) where $\bar{\lambda}$, k_1 are defined by:

$$(35.1)\quad \bar{\lambda} = -\frac{\varepsilon}{\pi a \nu}\int_0^{2\pi} f(a \cos\phi, -a\nu \sin\phi) \sin\phi \, d\phi$$

$$(35.2)\quad k_1 = \frac{\varepsilon}{\pi a}\int_0^{2\pi} f(a \cos\phi), -a\nu \sin\phi) \cos\phi \, d\phi.$$

The equations for δ, ω as demanded by a linear system are

$$\frac{da}{dt} = -\delta a, \ \frac{d\psi}{dt} = \omega$$

and they are precisely those of the first approximation.

The formal process just described will be referred to as the principle of linearization.

36. What is the physical significance of linearization? To answer the question we will have to have recourse to an electrical system.

We first recall certain concepts familiar in electrical engineering. Let

$$(36.1)\quad e(t) = E \cos \omega t, \ i(t) = I \cos (\omega t - \alpha)$$

be a harmonic voltage and harmonic current in a given circuit. The angle α is the phase-lag of $i(t)$, and $\cos \alpha$ is the power-factor of the system. The complex representatives of e, i are

$$(36.2)\quad \vec{e}(t) = Ee^{j\omega t}, \ \vec{i}(t) = Ie^{j(\omega t - \alpha)}, \ j = \sqrt{-1}.$$

Denoting temporarily by $\hat{x}$ the conjugate of any quantity x, if $T = \frac{2\pi}{\omega}$ is the period of the oscillations, then

$$(36.3)\quad \frac{1}{T}\int_{\delta}^{T} \vec{e}(t)\, \hat{\vec{i}}(t)dt = P_a - jP_r$$

where P_a, the mean power of the system, (measured or effective power) is known as the active power, and P_r as the reactive power. We also have

$$(36.4)\qquad \begin{cases} P_a = \frac{1}{T}\int_0^T e(t)i(t)dt \\ P_r = \frac{1}{T}\int_0^T e(t)i(t - \frac{T}{4}). \end{cases}$$

These last expressions may serve to define P_a, P_r for any periodic e, i.

Now (34.1) represents the motion of a particle subjected to the force $-kx-\varepsilon f$. Assuming the motion harmonic and of period T, the mean power consumed or active power will be

$$(36.5)\qquad P_a = \frac{1}{T}\int_0^T (kx+\varepsilon f(x, x'))x'dt$$

By an obvious analogy we may introduce here also a reactive power

$$(36.6)\qquad P_r = \frac{1}{T}\int_0^T (kx+\varepsilon F(x, x'))x'(t - \frac{T}{4})dt.$$

If we impose upon the linear system (34.8) the condition that its active and reactive powers be P_a, P_r, to within terms in ε^2, we obtain precisely the values given by (35.1),(35.2).

37. Another physical interpretation may also be obtained quite directly as follows. Substitute the harmonic oscillation $x = a \cos (\nu t+\theta)$ in the relations (34.9), (34.10). For this harmonic oscillation the equivalent linear force F_1 will be likewise harmonic with frequency ν. Let ϕ_1, I_1 denote the phase and

amplitude of F_1 so that

$$F_1 = I_1 \cos (\nu t+\phi_1).$$

The non-linear force will be periodic but with various harmonics whose frequencies will be multiples of ν. Let the fundamental harmonic be $I \cos (\nu t+\phi)$. If we equate the amplitude and phase of F_1 and of this fundamental harmonic:

$$I_1 = I, \quad \phi_1 = \phi$$

then we obtain relations which yield again (35.1), (35.2). In point of fact in expanded form the linear force will be

$$k_1 a \cos (\nu t+\theta) - \nu\bar{\lambda} a \sin (\nu t+\theta)$$

while the fundamental harmonic of the non-linear force is

$$\left\{\frac{1}{\pi}\int_0^{2\pi} f(a \cos \tau, -a\nu \sin \tau) \cos \tau \, d\tau\right\} \cos (\nu t+\theta)$$

$$+ \left\{\frac{1}{\pi}\int_0^{2\pi} f(a \cos \tau, -a\nu \sin \tau) \sin \tau \, d\tau\right\} \sin (\nu t+\theta).$$

If we equate the two it is but a step to (35.1),(35.2). The process just described for obtaining k, $\bar{\lambda}$ will be referred to as the principle of harmonic balance. There is no difficulty in showing that it is in fact equivalent to the first procedure for deriving (35.1),(35.2).

It is important to observe that there is no reason whatever to derive the differential equation for the oscillations before linearizing the system. Indeed, in many cases (especially for more or less complicated oscillatory systems) it may actually be more convenient to dispense with the formation of the differential equation, or to form it only afterwards and to linearize the system directly from the data. The basic fact is that we are dealing with systems which do not differ too much from harmonic systems.

38. We will now consider a few examples.

(38.1) Example 1. Suppose that we have a particle subjected to a non-linear spring whose effect is described by $F = f(x)$. Then for a harmonic oscillation $x = a \cos (\nu t+\theta)$ the fundamental harmonic in F will be

$$\left\{\frac{1}{\pi}\int_0^{2\pi} f(a \cos \tau) \cos \tau \, d\tau\right\} \cos (\nu t+\theta).$$

Therefore by the principle of harmonic balance we may replace the non-linear spring by a linear spring whose spring constant is

$$k(a) = \frac{1}{\pi a}\int_0^{2\pi} f(a \cos \phi) \cos \phi \, d\phi \ .$$

(38.2) Example 2. Consider a circuit with an iron core and let Φ, i be the flux and current with

$$\Phi = f(i) \tag{38.3}$$

as the relation between them. If the current is harmonic:

$$i = a \cos (\nu t+\theta), \tag{38.4}$$

then the fundamental harmonic of the flux will be

$$\left\{\frac{1}{\pi}\int_0^{2\pi} f(a \cos \phi) \cos \phi \, d\phi\right\} \cos (\nu t+\theta).$$

Therefore, by the principle of harmonic balance we may replace (38.3) by the equivalent linear relation $\Phi = L_e i$ where

$$L_e = \frac{1}{\pi a}\int_0^{2\pi} f(a \cos \phi) \cos \phi \, d\phi.$$

By analogy with linear circuits, we will call L_e the equivalent coefficient of self-induction.

(38.5) Example 3. Suppose that we have an electrical series circuit with the same inductor as in the preceding example and in addition a linear inductor with self-induction coefficient L and a capacity C. By linearization we obtain an equivalent system with coefficient of self-induction $L + L_e$ and capacity C. Therefore the frequency is approximately

$$\omega = \frac{1}{\sqrt{(L+L_e)C}} = \frac{1}{\sqrt{LC}}\left(1 - \frac{L_e}{2L}\right).$$

(38.6) Example 4. Consider an electrical circuit with a non-linear element N and characteristic relation $e = -F(i)$. If the current is again given by (38.4) then the fundamental harmonic of the voltage will be

$$(38.7) \quad \left\{ - \frac{1}{\pi a}\int_0^{2\pi} F(a \cos \phi) \cos \phi \, d\phi \right\} \cos (\nu t+\theta),$$

and so the non-linear element N may be replaced by a linear element with characteristic relation

(38.8) $e = -R_e i,\ R_e = \frac{1}{\pi a}\int_0^{2\pi} F(a\cos\phi)\cos\phi\, d\phi.$

This assumes, of course, that the circuit is such that the oscillations are nearly harmonic.

If R_e is positive then the circuit acts as an ohmic resistance and absorbs mean power to the amount of

$$\frac{R_e a^2}{2}.$$

If on the contrary R_e is negative then the circuit generates power to the same amount in absolute value. The system is then said to have the characteristic of a generator.

VI. APPLICATION OF SYMBOLIC METHODS TO LINEARIZATION

39. Let us introduce the linear operator j whose domain are the sines and cosines, and which is defined by

$$j \sin \omega t = \cos \omega t,\quad j \cos \omega t = -\sin \omega t$$

so that in particular

(39.1) $$j^2 = -1.$$

It is a consequence of (39.1) that j has the characteristic values $\pm i$ ($i = \sqrt{-1}$) and related characteristic functions $e^{i\omega t}$, $e^{-i\omega t}$. Moreover if $\Phi(z)$ is a function of the complex variable z and $\Phi(i) = A+iB$, A and B real, then $\Phi(j) = A+jB$. Furthermore if $A+iB = re^{i\alpha}$ then $\Phi(j) = re^{j\alpha}$. We also have

(39.2) $\Phi(j) \cdot a \cos(\omega t+\phi) = ra \cdot \cos(\omega t+\phi+\alpha)$

(39.3) $a \cos(\omega t+\phi) = e^{j\psi} \cdot a \cos \omega t$

whose effect is obvious. If f(t) is harmonic and of period $\frac{2\pi}{\omega}$ then

$$\frac{df(t)}{dt} = \omega j f(t)$$

or in operator form

(39.4) $$\frac{d}{dt} = \omega j, \quad j = \frac{1}{\omega}\frac{d}{dt} .$$

40. Consider now a linear conductor to whose terminals is applied a harmonic (sinusoidal) voltage of frequency ω. Kirchoff's law will yield a linear differential equation with constant coefficients for the current i(t). In view of (39.4) if there is a harmonic solution then it will satisfy a relation

(40.1) $$Z(j\omega)i = -e.$$

The operator $Z(j\omega)$ is known as the impedance of the conductor. Notice that we have for the complex current and voltage $\vec{i}$, $\vec{e}$:

$$Z(j\omega)\vec{i} = -\vec{e}.$$

It is often convenient to introduce also the inverse operator

(40.2) $$A(j\omega) = \frac{1}{Z(j\omega)}$$

known as the admittance of the conductor. More generally

let Σ be a network with two terminals and let e, i have the same meaning. Kirchoff's laws yield then the similar relations and so Σ has an impedance and an admittance.

As a simple example if an inductor L, a capacity C, and a resistance R are connected in series, the impedance is

$$Z = Lj\omega + R + \frac{1}{Cj\omega} \tag{40.3}$$

while if they are in parallel the admittance is

$$A = \frac{1}{Lj\omega} + \frac{1}{R} + Cj\omega. \tag{40.4}$$

The concepts of admittance and impedance which proved so important in the theory of alternating currents have been extended in recent years to other branches of physics notably to mechanical and acoustical systems.

Consider for instance the motion of a particle governed by the equation

$$m\frac{d^2x}{dt} + \lambda\frac{dx}{dt} + kx = f(t), \tag{40.5}$$

or with the velocity $v = \frac{dx}{dt}$ as unknown, by an equation

$$m\frac{dv}{dt} + \lambda v + k\int^t vdt = f(t). \tag{40.6}$$

Consider on the other hand an electrical circuit governed by the relation

$$L\frac{di}{dt} + Ri + \frac{1}{C}\int^t idt = e \tag{40.7}$$

where L, R, C, e, have the usual interpretation. In view of the complete formal identity between the two differential equations (40.6), (40.7) one may, with Pierre Curie (see his "Works", p. 164, 1891) establish the following analogies:

Mechanical Oscillations	Electrical Oscillations
Displacement x	Electrical Charge q
Velocity v	Current i
Force f	Voltage e
Mass m	Self-induction L
Friction Coefficient λ	Resistance R
Spring Constant k	Inverse of the Capacity C

This is in the main the electrical-mechanical analogy utilized in modern acoustics.

In connection with mechanical systems, one has frequent occasion to consider rotating systems. The basic differential equation for the possible oscillations of such a system with one degree of freedom will be of the form

$$J \frac{d^2\theta}{dt^2} + n \frac{d\theta}{dt} rc\theta = M. \tag{40.8}$$

It will be seen that it is obtained from (40.5) if x is replaced by the angular variable θ, the velocity by the angular velocity, the force f by the torque M, the mass by the moment of inertia J, the friction coefficient λ by the friction moment referred to the unit of velocity n, and finally the spring constant by the coefficient of hardness c. We thus have the following analogy between rotating mechanical systems and electrical systems:

Rotating Oscillations	Electrical Oscillations
Angular Displacement θ	Electrical Charge q
Angular Velocity $\frac{d\theta}{dt}$	Current i
Torque M	Voltage e
Moment of Inertia J	Self-induction L
Brake-torque n	Resistance R
Hardness c	Inverse Capacity $\frac{1}{C}$

Consider now the harmonic oscillations of these various systems.

For the electrical system (40.7) we have

$$(40.9) \qquad e = Zi, \; Z = Lj\omega + R + \frac{1}{Cj\omega} \,.$$

On the other hand by the operational method applied to (40.6) we obtain

$$(40.10) \qquad f = zv, \; z = mj\omega + \lambda + \frac{k}{j\omega} \,.$$

By analogy we introduce the mechanical impedance z and in association with it the mechanical admittance $y = \frac{1}{z}$.

41. Consider a linear electrical network without impressed voltages. If we examine the possible existence of harmonic self-oscillations of a given frequency ω, Kirchoff's laws yield a linear homogeneous algebraic system of equations with a determinant $\Delta(j\omega)$ rational in $j\omega$. A necessary condition will then be

$$(41.1) \qquad \Delta(j\omega) = 0.$$

The roots of (41.1) will specify the acceptable frequencies.

(41.2) Example 1. The net $\sum$ consists of a single

closed circuit with characters L, R, C. Then (41.1) becomes

$$(41.3) \qquad Z(j\omega) = Lj\omega + R + \frac{1}{j\omega C} = 0,$$

which is equivalent to

$$(41.4) \qquad R = 0, \omega = \frac{1}{\sqrt{LC}} .$$

Thus harmonic self-oscillation is only possible when the resistance R = 0, and then ω has the value indicated. These are, of course, well known facts.

(41.5) Example 2. Let the net Σ have two terminals and let there be impressed a harmonic voltage

$$e = E \cos (\omega t + \alpha)$$

at the terminals. The corresponding complex voltage is

$$\vec{e} = Ee^{j(\omega t + \alpha)}$$

and so the complex current i is defined by

$$Z(j\omega)\vec{i} = \vec{e}.$$

If we have $Z(j\omega) = |Z(j\omega)| \, e^{j\beta}$ then we find

$$(41.6) \qquad \vec{i} = \frac{Ee^{j(\omega t + \alpha - \beta)}}{|Z(j\omega)|} ,$$

and hence

$$(41.7) \qquad i = I \cos (\omega t + \alpha - \beta), \; I = \frac{|E|}{|Z(j\omega)|} .$$

One recognizes here the well known relations of alternating current theory.

(41.8) Example 3. Σ consists of two circuits with characters (L, R, C), (L_1, R_1, C_1) and coefficient of mutual induction M. If $\vec{I}$, $\vec{I}_1$ are the (complex) currents then Kirchoff's laws yield

$$(41.9)\qquad \begin{aligned} &(Lj\omega + \frac{1}{Cj\omega})\vec{I} - Mj\omega\vec{I}_1 = 0, \\ &(L_1 j\omega + \frac{1}{C_1 j\omega})\vec{I}_1 - Mj\omega\vec{I} = 0, \end{aligned}$$

and so

$$(41.10)\quad \Delta(j\omega) = (Lj\omega + \frac{1}{Cj\omega})(L_1 j_1\omega + \frac{1}{C_1 j\omega}) + M^2\omega^2 = 0.$$

Setting

$$(41.11)\qquad \nu = \frac{1}{\sqrt{LC}},\quad \nu_1 = \frac{1}{\sqrt{L_1C_1}},\quad q = \frac{M}{\sqrt{L_1L_2}},$$

the roots ω_1^2, ω_2^2 of (41.10) (considered as a quadratic in ω^2) are given by

$$(41.12)\qquad \left.\begin{aligned} \omega_1^2 \\ \omega_1^2 \end{aligned}\right\} = \frac{\nu^2 + \nu_1^2 \pm \sqrt{(\nu^2 - \nu_1^2)^2 + 4q^2\nu^2\nu_1^2}}{2(1 - q^2)}$$

provided that $q^2 \neq 1$. The admissable frequencies will be ω_1, ω_2 if they are real.

42. By combining the preceding developments with linearization, their range of application may profitably be extended to nets with non-linear elements, as we shall now show.

Returning to the first example the system may undergo an oscillation of the form

(42.1) $$i = Ae^{-\delta t} \cos (\omega t+\phi),$$

where

(42.2) $$\delta = \frac{R}{2L}, \quad \omega = \frac{1}{\sqrt{LC}}\sqrt{1 - \frac{R^2C}{4L}}.$$

Assuming now R small and considering $\frac{R}{2L}$ as of the first order of smallness we will have as a first approximation $\omega = \frac{1}{\sqrt{LC}}$ and thus nearly harmonic oscillations of frequency ω. To neutralize dissipation let us insert in the circuit a non-linear element N with characteristic $e = -F(i)$, such that the "instantaneous" resistance $F'(i)$ is of the order of R and not always positive in the range under consideration. Referring to (38.8) we replace N by an equivalent linear element with characteristic $e = -R_e i$, where

(42.3) $$R_e = \frac{1}{\pi a}\int_0^{2\pi} F(a \cos \phi)\cos \phi \, d\phi$$
$$= \frac{1}{\pi}\int_0^{2\pi} F'(a \cos \phi)\sin^2 \phi \, d\phi$$

where a is the amplitude of i. Clearly R_e is of the same order as $F'(i)$ and hence of the same order as R. The equivalent linear system is a circuit with characteristics $(L, R+R_e, C)$ and so this time as in (41.2):

(42.4) $$L\omega - \frac{1}{\omega C} = 0, \quad R + R_e = 0.$$

A stationary oscillation $i = a \cos (\omega t+\theta)$ will be determined in the first approximation by

(42.5) $$R_e(a) = -R, \quad \omega = \frac{1}{\sqrt{LC}}.$$

Its occurrence is clearly impossible unless $R_e(a)$ is not always positive, i. e., unless $F(i)$ has "falling" parts.

Consider more generally a linear net Σ with impedance $Z(j\omega)$ short-circuited on the same non-linear element N as above. This time linearization yields

$$(42.6) \qquad Z(j\omega) + R_e = 0.$$

Hence if $Z(j\omega) = X(\omega) + jY(\omega)$ we will have for a stationary oscillation $i = a \cos(\omega t+\theta)$:

$$(42.7) \qquad Y(\omega) = 0$$

$$(42.8) \qquad R_e(a) = -X(\omega).$$

The first relation determines ω, and then the second the amplitude a.

If the characteristic of the non-linear element N were of the form $i = f(e)$ we would proceed similarly with impedance replaced by admittance, and resistance by conductance.

43. The operator method duly generalized may be applied to non-stationary oscillations. Generally speaking in a linear system a non-stationary oscillation is of the exponential-harmonic type:

$$(43.1) \qquad x = Ae^{-\delta t} \cos(\omega t+\phi).$$

It satisfies the relation

$$(43.2) \qquad \frac{df(t)}{dt} = (-\delta + j\omega)f(t).$$

Hence all our arguments may be extended to exponential-harmonic oscillations provided that $j\omega$ is replaced everywhere by the operator $p = -\delta + j\omega$. For example if a network has for characteristic equation (41.1) then its non-stationary (understood exponential-harmonic) oscillations are governed by

$$(43.3) \qquad \Delta(p) = 0.$$

Thus for the same network Σ as before with non-linear element N we will have

$$(43.4) \qquad Z(p) + R_e(a) = 0, \; p = -\delta + j\omega.$$

Having determined by this relation δ and ω as functions of a, the elements of (43.1) will be given by the equations of the first approximation

$$(43.5) \qquad \frac{da}{dt} = -\delta a, \; \frac{d\psi}{dt} = \omega, \; \psi = \omega t + \phi.$$

For in the first approximation the solution assumes the form (43.1) with $a = Ae^{-\delta t}$, $\psi = \omega t + \phi$ and this implies (43.5).

Suppose in particular that Σ consists merely of the series circuit (L, R, C). Then $Z(p) = Lp + R + \frac{1}{Cp}$ and (43.4) reads

$$Lp + (R + R_e) + \frac{1}{Cp} = 0.$$

Hence here

$$(43.6) \qquad \delta(a) = \frac{R + R_e(a)}{2L}, \quad \omega = \frac{1}{\sqrt{LC}}.$$

There will be self-excitation in the system if (43.1) does not die down when a is neai zero, i. e. if $\delta(o) < o$

or if $R_e(o)<R^*$, $R^* = -R$. Thus R^* (here $-R$) is a critical equivalent resistance for N such that below it the system is self-oscillatory, above it is not.

VII. MULTIPLY PERIODIC SYSTEMS

44. Up to the present the oscillations under consideration have been taken so to speak one at a time. If the system is linear and several frequencies are admissible, say if the characteristic equation (41.1) has the roots $\omega_1, \ldots, \omega_n$, then there are possible stationary oscillations

$$i_h = a_h \cos(\omega t + \phi_h), \tag{44.1}$$

and so by the principle of superposition (for linear systems) there is a stationary solution

$$i = a_1 \cos(\omega_1 t + \phi) + \ldots + a_n \cos(\omega_h t + \phi_n). \tag{44.2}$$

It is fairly clear that this principle may not be applied to a linear system equivalent to a given non-linear system. Under certain conditions (reasonable smallness of suitable parameters) some progress may still be made, as we shall now show. For simplicity we limit the discussion to the case of two oscillations. Two distinct situations will arise according to the presence or absence of resonance.

45. Consider then a linear system Σ with two terminals across which there is connected a non-linear element N whose characteristic we write as

$$(45.1) \qquad e = -F(i) = -\varepsilon f(i)$$

where ε will serve to gauge the deviation from linearity. We will suppose also that f is a polynomial. In practice this is a rather mild restriction if f is continuous (the usual case), since it may then be arbitrarily and uniformly approximated by a polynomial over any closed interval.

Let $Z(j\omega)$ be the impedance of the network and ω_{10}, ω_{20} its fundamental (natural) frequencies. We will then have

$$(45.2) \qquad Z(j\omega) = (\omega_{10}^2 - \omega^2)(\omega_{20}^2 - \omega^2)g(j\omega)$$

where $g(j\omega_{ho}) \neq o$, $h = 1, 2$.

Assuming now ε small let the non-linear system admit the oscillations represented in the first approximation by

$$(45.3) \qquad i_h = a_h \cos(\omega t + \phi_h),$$

where a_h, ϕ_h are arbitrary constants and where $\omega_h^2 = \omega_{ho}^2 + \varepsilon\zeta_h$, to the order ε^2. Set now

$$(45.4) \quad i = i_1 + i_2 = a_1 \cos(\omega t + \phi_1) + a_2 \cos(\omega t + \phi_2).$$

The resulting voltage in N is

$$(45.5) \; e = -\varepsilon f(i) = -\varepsilon f(a_1 \cos(\omega_1 t + \phi_1) + a_2 \cos(\omega_2 t + \phi_2)).$$

Consider now the double Fourier expansion with respect to ϕ_1, ϕ_2:

(45.6) $f(a_1\cos\phi_1, a_2\cos\phi_2) = \sum A_{mn}\cos(m\phi_1+n\phi_2).$

Since f is a polynomial the sum is finite. We have now

(45.7) $e = -\varepsilon\sum A_{mn}\cos((m\omega_1+n\omega_2)t+m\phi_1+n\phi_2).$

The voltage may be considered as a sum of voltages

(45.8) $e_{mn} = -\varepsilon A_{mn}\cos((m\omega_1+n\omega_2)t+m\phi_1+n\phi_2)$

applied to the linear network of impedance $Z(j\omega)$. This is where we must distinguish between two possible situations.

46. NON RESONANT SYSTEM. Suppose first that none of the expressions

(46.1) $(m\pm1)\omega_1 + n\omega_2,\ m\omega_1 + (n\pm1)\omega_2$

except those corresponding to $m = 1$, $n = 0$, and $m = 0$, $n = 1$ are of order at least ε. This will certainly hold if none of the frequencies

(46.2) $(m\pm1)\omega_{10} + n\omega_{20},\ m\omega_{10} + (n\pm1)\omega_{20}$

other than those corresponding to $(m, n) = (1,0), (0,1)$ are zero. Referring then to (41.5) it is seen that under the circumstances for $(m,n) \neq (1,0), (0,1)$, e_{mn} induces in Σ a current

(46.3) $$\begin{cases} i_{mn} = \dfrac{-\varepsilon A_{mn}\cos((m\omega_1+n\omega_2)t+m\phi_1+n\phi_2-\beta_{mn})}{|Z(j(m\omega_1+n\omega_2)|} \\ Z(j(m\omega_1+n\omega_2)) = |Z|e^{j\beta_{mn}}. \end{cases}$$

The amplitude of i_{mn} is

$$I_{mn} = \frac{|\epsilon A_{mn}|}{|Z(j(m\omega_1+n\omega_2)|} . \tag{46.4}$$

Under the assumption $m\omega_1+n\omega_2$ differs by a finite quantity from ω_{10}, ω_{20} and so I_{mn} is of order ϵ.

On the contrary for instance $Z(j\omega_1)$ is of order ϵ and so I_{10}, and similarly I_{01}, is finite. Thus to the order ϵ the current generated will be

$$\begin{aligned} i = i_{10}+ i_{01} &= I_{10}\cos (\omega_1 t+\phi_1-\beta_{10}) \\ &+ I_{01}\cos (\omega_2 t+\phi_2-\beta_{01}). \end{aligned} \tag{46.5}$$

This must be the same as (45.4) to within the order ϵ. The identification yields first $\beta_{10} = \beta_{01} = 0$ to the order ϵ, a condition already satisfied. Then we must have $a_1 = I_{10}$, $a_2 = I_{01}$ and so

$$a_1 = \frac{-\epsilon A_{10}}{|Z(j\omega_1)|}, \quad a_2 = \frac{-\epsilon A_{01}}{|Z(j\omega_2)|} . \tag{46.6}$$

From this follows readily that if we set

$$R_{eh} = \frac{1}{2\pi^2 a_h} \int_0^{2\pi}\int_0^{2\pi} F(a_1\cos \phi_1+a_2\cos \phi_2)\cos \phi_h d\phi, \quad (h=1,2) \tag{46.7}$$

then

$$e = -(R_{e1}i_1+R_{e2}i_2). \tag{46.8}$$

Thus in the first approximation the non-linear characteristic $e = -\varepsilon f(i) = -\varepsilon f(i_1+i_2)$ may be replaced by the linear characteristic (46.8). We may therefore interpret the replacement of the non-linear element N by an equivalent element with the characteristic just written as a linearization of the system when there is no resonance.

As in the case of a single oscillation we may write down here also a refined first approximation for i, representing it to within the order ε^2, and it will be, neglecting no i_{mn} term:

$$(46.9) \quad i = \sum i_{mn} = a_1 \cos (\omega t+\phi_1)+a_2 \cos (\omega_2 t+\phi_2) -\varepsilon \sum \frac{A_{mn} \cos (m\omega_1 +n\omega_2+\theta_{mn})}{|Zj(m\omega_1 +n\omega_2)|}$$

$$(46.10) \qquad \theta_{mn} = m\phi_1+n\phi_2-\beta_{mn}.$$

47. <u>RESONANT SYSTEM</u>. Suppose now that one of the frequencies (46.1) is zero, or $\omega_{20} = \frac{r}{s}\omega_{10}$, where r, s are relatively prime and $\frac{r}{s}$ is one of the fractions $\frac{-m\pm 1}{n}$, $\frac{-m}{n\pm 1}$. It is clear that we may assume without restriction $\omega_2 > \omega_1$, or $r > s$. This time we will have a finite Fourier sum:

$$(47.1) \qquad f(a_1 \cos (\omega_1 t+\phi) + a_2 \cos (\omega_2 t+\phi_2)) = \sum A_m \cos (\tfrac{m}{s}\omega_1 t+\theta_m).$$

If we set

$$(47.2) \qquad e_m = -\varepsilon A_m \cos (\tfrac{m}{s}\omega_1 t+\theta_m)$$

then we have

$$(47.3) \qquad e = \sum e_m = -\varepsilon \sum A_m \cos \left(\frac{m}{s}\omega_1 t + \theta_m\right).$$

The current i_m induced by e_m in $\sum$ is

$$(47.4) \qquad i_m = \frac{-\varepsilon A_m \cos \left(\frac{m}{s}\omega_1 t + \theta_m - \beta_m\right)}{|Z(j\frac{m}{s}\omega_1)|}$$

$$(47.5) \qquad Z(j\frac{m}{s}\omega_1) = |Z| e^{j\beta m}.$$

The amplitude of i_m is

$$(47.6) \qquad I_m = \frac{|\varepsilon A_m|}{|Z(j\frac{m}{s}\omega_1)|}$$

and it is finite for $m = r, s$, of order ε otherwise. Thus

$$i = i_r + i_s = I_s \cos (\omega_1 t + \theta_s - \beta_s) + I_r \cos (\omega_2 t + \theta_r - \beta_r).$$

The identification with $i = i_1 + i_2$ yields here

$$a_1 = I_s, \; a_2 = I_r, \; \theta_s - \beta_s = \phi_1, \; \theta_r - \beta_r = \phi_2.$$

The linearization assumes this time the form

$$e = -(Z_1 i_1 + Z_2 i_2)$$

where Z_1, Z_2 are (complex) impedances whose computation offers no particular difficulty.

The "refined" first approximation for i, or approximation to the order ε^2 is

$$(47.7) \qquad i = a_1 \cos(\omega_1 t + \phi_1) + a_2 \cos(\tfrac{r}{s}\omega_1 t + \phi_2)$$

$$-\varepsilon \sum_{m \neq r,s} \frac{A_m \cos(\frac{m}{s}\omega_1 t + \theta_m - \beta_m)}{|Z(j\frac{m}{s}\omega_1)|} .$$

In addition to the harmonics ω_1, ω_2 it will contain others of the form $\frac{m}{s}\omega_1$, which may be of smaller frequency than ω_1. Thus we have here so-called demultiplication of frequency, a property of considerable practical importance.

VIII. INFLUENCE OF PERIODIC DISTURBANCES.

48. Up to the present we have concentrated upon isolated systems, not subjected to any exterior disturbances. As an example of a non-isolated system we will discuss the equation

$$(48.1) \qquad m\frac{d^2x}{dt^2} + kx = \varepsilon f(t, x, \frac{dx}{dt}),$$

where k, m are positive, ε is small and

$$(48.2) \qquad f(t, x, \frac{dx}{dt}) = f_o(x, \frac{dx}{dt}) + \sum (f_n^*(x, \frac{dx}{dt}) \cos \lambda_n t$$

$$+ f_n^{**}(x, \frac{dx}{dt}) \sin \lambda_n t),$$

where the sum is finite and f_o, f_n^*, f_n^{**} are polynomials.

The mechanical interpretation of (48.1) is obvious. As equivalent electrical system we may choose a series circuit with current $i = \frac{dx}{dt}$, inductor m, capacity $\frac{1}{k}$, condenser charge x, and non-linear element N whose characteristic is

$$(48.3)\qquad e = \varepsilon f(t,x,i).$$

Since for ε small the system is quasi-harmonic we shall apply the general concept of linearization. For $\varepsilon = 0$ we may choose

$$(48.4)\qquad x = a\sin(\omega_o t+\phi),$$

$$(48.5)\qquad i = \frac{dx}{dt} = a\omega_o\cos(\omega_o t+\phi),$$

$$(48.6)\qquad \omega_o = \sqrt{\frac{k}{m}}\,.$$

For ε small but $\neq 0$ we will consider the above formulas as approximations and substitute them in (48.3). We have then

$$(48.7)\qquad e = \varepsilon f(t,\ a\sin(\omega_o t+\phi),\ a\omega_o\cos(\omega_o t+\phi)).$$

Since f_o, f_n^*, f_n^{**} are polynomials we have finite Fourier sums:

$$(48.8)\qquad f_o(a\sin\psi, a\omega_o\cos\psi) = \sum(f_k(a)\cos k\psi + g_k(a)\sin k\psi)$$

and similarly for f_n^*, f_n^{**} with f_{nk}^*, f_{nk}^{**}, and g_{nk}^*, g_{nk}^{**} as the coefficients. Hence

$$(48.9)\qquad e = \sum(f_k\cos k(\omega_o t+\phi) + g_k\sin k(\omega_o t+\phi)) + \{\ldots\ldots\}$$

the unwritten terms containing sines and cosines of the angles $k\omega_o \pm \lambda_n$, with the ranges of k and of the λ_n

finite. Here again we must distinguish between resonance and non-resonance accordingly as $k\omega_0 \pm \lambda_n = \omega_0$, does or does not hold for some (k, λ_n).

49. NON-RESONANT SYSTEM. This is the case where no frequency $k\omega_0 \pm \lambda_n$ is ω_0 itself, or where $(k-1)\omega_0 \neq \pm\lambda_n$, whatever k, λ_n in their ranges. Then the only harmonic of frequency ω_0 in e is

(49.1) $$e_1 = \varepsilon(f_1(a)\cos(\omega_0 t+\phi) + g_1(a)\sin(\omega_0 t+\phi)).$$

In view of (48.6): $\vec{e}_1 = Z_e\vec{i}$, where

(49.2) $$Z_e = \frac{\varepsilon}{\omega_0 a}(f_1(a)-jg_1(a)).$$

By the basic principle governing linearization we replace the non-linear element N by an equivalent linear element with characteristic

$$e = Z_e i.$$

The equivalent linear system has then the impedance $m\omega j + \frac{k}{\omega\delta} - Z_e$ and its characteristic equation is $z(p) = Z_e$, or explicitly (see 43):

$$mp + \frac{k}{p} = \frac{\varepsilon}{\omega_0 a}(f_1(a)-jg_1(a)), p = -\delta+j\omega.$$

Therefore:

$$m(-\delta+\omega j) + \frac{k}{-\delta+\omega j} = \frac{\varepsilon}{\omega_0 a}(f_1(a)-jg_1(a)),$$

and this yields to within the order ε^2

(49.3) $$\begin{cases} \delta = \dfrac{-\varepsilon}{2m\omega_0 a} f_1(a) \\ \omega = \omega_0 - \dfrac{-\varepsilon}{2m\omega_0 a} g_1(a). \end{cases}$$

If we combine with the equations of the first approximation (see 43):

$$\frac{da}{dt} = -\delta a, \ \frac{d\psi}{dt} = \omega,$$

we may replace the latter by

$$(49.4) \quad \begin{cases} \frac{da}{dt} = \frac{\varepsilon}{2m\omega_0} f_1 \ (a), \\ \frac{d\psi}{dt} = \omega_0 - \frac{\varepsilon}{2m\omega_0 a} g_1(a) = \omega(a), \\ \psi = \omega_0 t+\phi. \end{cases}$$

The related first approximation for x is.

$$(49.5) \qquad x = a \sin \psi.$$

Introduce now the expressions

$$(49.6) \qquad \bar{\lambda} = \frac{-\varepsilon}{a\omega_0} f_1(a), \ k_e^* = \frac{-\varepsilon}{a} g_1(a).$$

Since $\omega_0^2 = \frac{k}{m}$, we have to within the order ε^2:

$$\omega^2 = \frac{k + k_e}{m} \ ,$$

and so (49.4) may be replaced by

$$(49.7) \quad \begin{cases} \frac{da}{dt} = - \frac{\bar{\lambda}}{2m} a, \\ \frac{d\psi}{dt} = \sqrt{\frac{k + k_e^*}{m}}. \end{cases}$$

From (49.5), (49.6), (49.7) we deduce to within the

order ε^2 the relation

$$m\frac{d^2x}{dt^2} + \bar{\lambda}\frac{dx}{dt} + (k + k_e^*)x = 0. \tag{49.8}$$

Notice that (49.8) depends solely upon the term f_o of f. Since

$$f_o(x,\frac{dx}{dt}) = \lim_{T\to+\infty} \frac{1}{T}\int_0^T f(\tau,x,\frac{dx}{dt})\,d\tau,$$

the linearization and associated first approximation may be obtained by applying the averaging process to f (averaging as to t, as if x, $\frac{dx}{dt}$ were independent variables), and replacing f by the resulting function f_o.

To sum up then: as regards the first approximation and related linearization we may replace f by $f_o(x,\frac{dx}{dt})$. Since this last function does not contain t explicitly, we have a situation already considered. We merely recall these properties:

(49.9) The stationary amplitudes $\neq 0$ are the solutions of $\bar{\lambda}(a) = 0$. If a_o is such a solution then the corresponding oscillation is stable whenever $\frac{d\bar{\lambda}}{da_o} > 0$, and unstable otherwise. This assumes, of course, that the derivative $\neq 0$ at a_o.

(49.10) Self-excitation occurs when and only when $\bar{\lambda}(0) < 0$.

(50) By way of example let us apply the preceding results to van der Pol's equation with a forced oscillation:

$$\frac{d^2y}{dt^2} - \varepsilon(1-y^2)\frac{dy}{dt} + y = E\sin\alpha t, \tag{50.1}$$

where as usual $\varepsilon > 0$. To reduce (50.1) to the form (48.1) set

(50.2) $$y = x + b \sin \alpha t, \quad b = \frac{E}{1-\alpha^2} .$$

Then x satisfies

(50.3) $$\frac{d^2x}{dt^2} + x = \varepsilon(1-(x + b \sin \alpha t)^2)(\frac{dx}{dt} + b\alpha \cos \alpha t).$$

Here

$$m = k = 1, \quad f_o(x,\frac{dx}{dt}) = (1 - \frac{b^2}{2} - x^2) \frac{dx}{dt} ,$$

$$\bar{\lambda} = (1 - \frac{a^2}{4} - \frac{b^2}{2}).$$

Hence the first approximation is

$$x = a \sin (t+\phi)$$

(50.4) $$\phi = \text{const.}, \quad \frac{da}{dt} = \frac{\varepsilon}{2} (1 - \frac{a^2}{4} - \frac{b^2}{2}) .$$

Therefore there is self-excitation when and only when $b^2<2$ and there is a stable stationary amplitude $a = \sqrt{4-2b^2}$. The corresponding stationary solution of (50.1) is

(50.5) $$y = b \sin \alpha t + \sqrt{4-2b^2} \sin (t+\phi).$$

For $b^2>2$, $x = 0$ is stable and so

(50.6) $$y = b \sin \alpha t$$

is a stable forced oscillation for (50.1).

51. RESONANT SYSTEM. To simplify matters we will suppose f of the form $f(\alpha t, x, \frac{dx}{dt})$, where $f(\tau, u, v)$ is periodic in τ and of period 2π. The basic equation is then

$$m \frac{d^2x}{dt^2} + kx = \epsilon f(\alpha t, x, \frac{dx}{dt}). \tag{51.1}$$

We suppose now that

$$\omega_0 = \frac{r}{s} \alpha + \epsilon \Omega$$

where $\frac{r}{s}$ is an irreducible fraction. As usual we set

$$x = a \sin (\frac{r}{s}\alpha t + \phi) \tag{51.2}$$

and replace $F = \epsilon f$ by the equivalent linear force $F_1 = -k_e x - \bar{\lambda} \frac{dx}{dt}$. By identifying the fundamental harmonics of

$$\epsilon f(\alpha t, a \sin (\frac{r}{s}\alpha t + \phi) - \frac{ar\alpha}{s} \cos (\frac{r}{s}\alpha t + \phi))$$

and

$$-k_e a \sin (\frac{r}{s}\alpha t + \phi) - \bar{\lambda} \frac{ar}{s} \cos (\frac{r}{s}\alpha t + \phi)$$

we obtain to within the order ϵ^2:

$$k_e = \frac{-\epsilon}{\pi a} \int_0^{2\pi} f(s\tau - \frac{s}{r}\psi, \text{ a } \sin r\tau, a\omega_0 \cos r\tau) \sin r\tau d\tau$$

$$\lambda = \frac{-\epsilon}{\pi a \omega_0} \int_0^{2\pi} f(s\tau - \frac{r}{s}\phi, \text{ a } \sin r\tau, a\omega_0 \cos r\tau) \cos r\tau d\tau.$$

The equivalent linear system is thus

(51.3) $$m\frac{d^2x}{dt^2} + \bar{\lambda}\frac{dx}{dt} + (k + k_e)x = 0.$$

Thus here to within the order ε^2

(51.4) $$\delta = \frac{\bar{\lambda}}{2m}, \quad \omega = \sqrt{\frac{k + k_e}{m}} = \omega_0(1 + \frac{k_e}{2k}),$$

and the equations of the first approximation are (see 43):

(51.5) $$\frac{da}{dt} = \frac{-\bar{\lambda}}{2m}a, \qquad \frac{d\psi}{dt} = \frac{k + k_e}{m} = \omega_0(1 + \frac{k_e}{2k})$$

since $\frac{k}{m} = \omega_0^2$.

(51.6) It is to be observed that if $f(\tau,u,v)$ is a finite trigonometric sum of terms $\sin k\tau$, $\cos k\tau$, then unless (r,s) is in a certain very limited range the t term in $f(t,x,\frac{dx}{dt})$ does not influence k_e, $\bar{\lambda}$, and as regards the first approximation it may be suppressed. In that case we are back to a previous case where f is of the form $f(x,\frac{dx}{dt})$ (no periodic disturbance). Roughly speaking it means that the resonances that count occur within a limited range of values (r,s).

Notice in particular that if $\frac{r}{s} = \frac{1}{s}$, then the frequency will be very near ω_0 and hence very near $\frac{\alpha}{s}$. Thus the application of a disturbance of frequency α may induce an effect of frequency $\frac{\alpha}{s}$. This is known as subharmonic resonance or demultiplication, and the property has been extensively applied especially in radio technique.

In point of fact, not only will the frequency $\frac{\alpha}{s}$ appear but also certain multiples which depend upon the nature of the polynomial f.

(51.7) Let us apply the preceding consideration to (50.1), the van der Pol equation with harmonic disturbance. For $s > 3$ we find the same situation as in (50) and nothing is changed. For $s = 3$ we must replace the second relation of (50.3) by

$$(51.8) \qquad \frac{da}{dt} = \frac{\varepsilon}{2}\left(1 - \frac{a^2}{4} - \frac{b^2}{2} - ab\right).$$

Here again self-excitation in x arises only for $b^2 > 2$ with a limiting stationary amplitude

$$a = -2b + \sqrt{4 + 2b^2}.$$

The corresponding stationary solution of (50.1) is given in the first (not refined) approximation by

$$y = (-2b + \sqrt{4+2b^2})\sin\left(\frac{t}{s} + \phi\right) + b \sin \alpha t.$$

Notice that when $b^2 = \frac{E}{1-\alpha^2} < \frac{4}{7}$, (weak disturbance) then the subharmonic dominates the harmonic.

For $b > 2$ the situation is as in (50) and there is no subharmonic.

IX. COMPLEMENTS

52. We will first discuss a somewhat different manner of obtaining the higher approximations from the procedure indicated in Chapter IV. Consider then the differential equation

$$(52.1) \qquad \frac{d^2x}{dt^2} + \omega^2 x = \varepsilon f\left(t, x, \frac{dx}{dt}, \varepsilon\right)$$

where for ε sufficiently small we have a power series representation

$$f(t, x, \frac{dx}{dt}, \varepsilon) = \sum \varepsilon^n f_n(t, x, \frac{dx}{dt}) \tag{52.2}$$

in which f_n is a polynomial in $x, \frac{dx}{dt}$, $\sin t$, $\cos t$. Regarding ω we assume explicitly that it is a positive irrational number.

Introduce now new variables a, θ defined by the relations

$$x = a \sin \theta, \quad \frac{dx}{dt} = a\omega \cos \theta. \tag{52.3}$$

This enables us to replace (52.1) by the system

$$\begin{cases} \frac{da}{dt} = \varepsilon f(t, a \sin \theta, a\omega \cos \theta, \varepsilon) \cos \theta \\ \frac{d\theta}{dt} = \omega - \frac{\varepsilon}{\omega a} f(t, a \sin \theta, a\omega \cos \theta, \varepsilon) \sin \theta. \end{cases} \tag{52.4}$$

Under our assumptions we also have

$$\begin{cases} f_o(t, a \sin \theta, a\omega \cos \theta) \cos \theta = F(a) \\ \qquad + \sum_{m^2+n^2 \neq 0} L_{mn}(a) e^{jm\theta+nt} \\ f_o(t, a \sin \theta, a \cos \theta) \sin \theta = \Phi(a) \\ \qquad + \sum_{m^2+n^2 \neq 0} M_{mn} e^{j(m\theta+nt)} \end{cases} \tag{52.5}$$

where the sums are finite. Let now

$$(52.6)\quad \begin{cases} u(a,\ \theta,\ t) = \sum L_{mn} \dfrac{e^{j(m\theta+nt)}}{j(m\omega+n)} \\ v(a,\ \theta,\ t) = \sum M_{mn} \dfrac{e^{j(m\theta+nt)}}{j(m\omega+n)} \ . \end{cases}$$

We verify at once the relations

$$(52.7)\quad \begin{cases} \dfrac{\partial u}{\partial t} + \omega \dfrac{\partial u}{\partial \theta} = f_o \cos \theta - F(a) \\ \dfrac{\partial v}{\partial t} + \omega \dfrac{\partial v}{\partial \theta} = f_o \sin \theta - \Phi(a). \end{cases}$$

We introduce now in place of a, θ new variables a_1, θ_1 defined by

$$(52.8)\quad \begin{cases} a = a_1 + \frac{\varepsilon}{\omega} u(a_1,\ \theta_1,\ t) \\ \theta = \theta_1 - \dfrac{\varepsilon}{\omega a_1} v(a_1,\ \theta_1,\ t) = \theta_1 - \frac{\varepsilon}{\omega} w(a_1,\ \theta_1,\ t). \end{cases}$$

By substituting the expressions (52.8) for a, θ in (52.4) we find

$$(52.9)\quad \begin{cases} \dfrac{da_1}{dt} + \dfrac{\varepsilon}{\omega} \dfrac{\partial u(a_1,\ \theta_1,\ t)}{\partial a_1} \dfrac{da_1}{dt} + \dfrac{\partial u(a_1,\ \theta_1,\ t)}{\partial \theta_1} \dfrac{d\theta_1}{dt} \\ \quad + \dfrac{\partial u(a_1,\ \theta_1,\ t)}{\partial t} = \frac{\varepsilon}{\omega} f(t, a \sin \theta,\ a\omega \cos \theta,\ ) \cos\theta \\ \dfrac{\partial \theta_1}{\partial t} - \dfrac{\varepsilon}{\omega} \dfrac{\partial w(a_1,\theta_1,t)}{-\partial a_1} \dfrac{da_1}{dt} + \dfrac{\partial w(a_1,\theta_1,t)}{\partial \theta_1} \dfrac{d\theta_1}{dt} \\ \quad + \dfrac{\partial w(a_1,\theta_1,t)}{\partial t} = \omega - \dfrac{\varepsilon}{\omega} f(t, a \sin \theta, a\omega \cos \theta,\ \varepsilon) \\ \qquad \sin \theta. \end{cases}$$

Eliminating a, θ by means of (52.8) and solving for $\frac{da_1}{dt}$, $\frac{d\theta_1}{dt}$, we obtain after some simplifications:

$$(52.10)\begin{cases} \dfrac{da_1}{dt} = \dfrac{\varepsilon}{\omega} F(a_1) + \varepsilon^2 R(a_1, \theta_1, t, \varepsilon), \\ \dfrac{d\theta}{dt} = \omega - \dfrac{\varepsilon}{\omega a_1} \Phi(a_1) + \varepsilon^2 S(a_1, \theta_1, t, \varepsilon), \end{cases}$$

where for ε very small we have expansions:

$$R = \sum \varepsilon^n R_n(a_1, \theta_1, t), \qquad S = \sum \varepsilon^n S_n(a_1, \theta_1, t),$$

with R_n, S_n polynomials in cos θ_1, sin θ_1, cos t, sin t. In particular

$$R_o = F_1(a_1) + \sum_{m^2+n^2\neq 0} L^1_{mn} e^{j(m\theta_1+nt)}$$

$$S_o = \Phi_1(a_1) + \sum_{m^2+n^2\neq 0} M^1_{mn} e^{j(m\theta_1+nt)},$$

where the sums are finite. The same reasoning may now be repeated with f_o cos θ, f_o sin θ replaced by R_o, S_o, etc. The final result may be described as follows. For each n there may be written a system of differential equations in a_n, θ_n:

$$(52.11)_n \quad \begin{aligned} \frac{da_n}{dt} &= -F(a_n) + \varepsilon^2 F_1(a_n) + \ldots + \varepsilon^n F_{n-1}(a_n) \\ &\quad + \varepsilon^{n+1} R^{(n)}(a_n, \theta_n, t, \varepsilon) \\ \frac{d\theta_n}{dt} &= \omega - \frac{\varepsilon}{\omega} \Phi(a_n) + \varepsilon^2 \Phi_1(a_n) + \ldots \\ &\quad + \varepsilon^n \Phi_{n-1}(a_n) + \varepsilon^{n+1} S^{(n)}(a_n, \theta_n, t, \varepsilon). \end{aligned}$$

In these equations $R^{(n)}$, $S^{(n)}$ have the same properties as R, S. Furthermore $R^{(o)} = R$, $S^{(o)} = S$ and if

$$R^{(k)} = \sum \varepsilon^s R_s^{(k)}, \quad S^{(k)} = \sum \varepsilon^s S_s^{(k)}$$

then

$$R_o^{(k-1)} = F_k(a_k) + \sum L_{mn}^k e^{j(m\theta_k + nt)}$$

$$S_o^{(k-1)} = \Phi_k(a_k) + \sum M_{mn}^k e^{j(m\theta_k + nt)}.$$

We may also assume that if ε is so small that terms of order of ε^{n+1} may be neglected then we have the following relations for the nth approximation:

$$x = X_n(a_n, \theta_n, t, \varepsilon),$$

where

$$(52.12)_n \quad \begin{aligned} \frac{da_n}{dt} &= \frac{\varepsilon}{\omega} F(a_n) + \varepsilon^2 F_1(a_n) + \dots + \varepsilon^n F_{n-1}(a_n) \\ \frac{d\theta_n}{dt} &= \omega - \frac{\varepsilon}{\omega a_n} \Phi(a_n) + \varepsilon^2 \Phi_1(a_n) + \dots \\ &\quad + \varepsilon^n \Phi_{n-1}(a_n). \end{aligned}$$

The method just described for obtaining the successive approximations is very direct and lends itself rather well to an estimation of the error consequent upon neglecting certain terms.

Another observation to be made is that the system $(52.12)_n$ may be deduced from $(52.11)_{n-1}$ be replacing the latter by the constant term in its expression as a double Fourier series, then rejecting terms of order ε^{n+1}.

A last remark regarding the process just described, is that continued indefinitely it does yield formal solutions as power series in ε, but unfortunately as shown by Poincaré, the series are generally divergent. Thus they cannot be utilized directly to investigate the structural properties of the solutions.

53. Passing now to an entirely different type of considerations we will discuss the following problem: - What indications do the approximations provide regarding the exact solutions?

Taking first (52.12), dropping the index 1: we have the system

$$\frac{da}{dt} = \frac{\varepsilon}{\omega} F(a) \tag{53.1}$$
$$\frac{d\theta}{dt} = -\frac{\varepsilon}{\omega a} \Phi(a).$$

The corresponding first approximation (the earlier "refined" first approximation) is

$$x = a \sin \theta + \frac{\varepsilon}{\omega}\{u(a,\theta,t)\sin \theta - v(a,\theta,t)\cos \theta\}. \tag{53.2}$$

If we set

$$f_o(t, a \sin \theta, a\omega \cos \theta) = \sum f_{mn}(a)e^{j(m\theta+nt)} \tag{53.3}$$

then we find readily in place of (53.2)

$$x = a \sin \theta + \varepsilon \sum \frac{f_{mn}(a)\ e^{j(m\theta+nt)}}{\omega^2 - (m\omega+n)^2} \tag{53.4}$$
$$(n^2 + (m^2-1)^2 \neq 0).$$

Suppose now that $a^* \neq 0$ is a simple solution of

$$(53.5) \qquad F(a) = 0.$$

Then $F'(a^*) \neq 0$ and we will suppose explicitly

$$(53.6) \qquad F'(a^*) < 0.$$

The other case would be dealt with by replacing everywhere t by -t.

It follows from (53.5) that (53.1) has the solution

$$(53.7) \qquad a = a^*, \ \theta = \nu t + \psi, \ \nu = \omega - \frac{\varepsilon}{\omega a^*} \Phi(a^*),$$

where ψ is an arbitrary constant. The corresponding stationary solution given by (53.3) is explicitly:

$$(53.8) \quad x = a^* \sin(\nu t + \psi) + \varepsilon \sum \frac{f_{mn}(a^*) e^{jm\psi}}{\omega^2 - (m\omega + n)^2} e^{j(m\nu + n)t}.$$

Thus it is of the form

$$(53.9) \qquad x = z(t, \nu t)$$

where $z(\theta, \phi)$ is a continuous periodic function of θ, ϕ with period 2π in each and depends upon ε. In particular $z(t, \nu t)$ will be quasi-periodic for all irrational ν.

In view of (53.6) we see that every solution of (12) for which the initial value a is near enough to a^* will tend with $t \to +\infty$ to one of the stationary solutions of (53.7).

Thus we may assert that every approximate solution (53.4) whose initial values x, $\frac{dx}{dt}$ are near enough to the initial determinations of the approximate stationary solution will tend with $t \to +\infty$ to one of these stationary régimes.

One may prove the following result: The property just formulated for approximate solutions (representation by quasi-periodic functions of the form (53.9) and properties of stability) belong also to the exact solution of the differential equation (52.1), at least whenever ε is sufficiently small.

This important property shows that the investigation of any particular approximation (for instance the first) obtained by the methods which we have repeatedly discussed, has a meaning not merely for purposes of approximation but may serve likewise to give heuristic indications regarding the structural qualities of the exact solutions.

The proof of this theorem has been given at length in Mémoire No. 16 of the Bibliography. We will merely discuss here two special cases which will serve as a strong indication regarding the nature of the theorem.

54. Consider first the case where ω is not an integer and (53.5) has the solution 0 with

$$F'(0) < 0. \tag{54.1}$$

We have thus $f_{mn}(0) = 0$ for $m \neq 0$. Hence the corresponding stationary solution (53.8) assumes the form

$$x = \varepsilon \sum_n \frac{f_n e^{jnt}}{\omega^2 - n^2}, \tag{54.2}$$

where

$$f_n = f_{n,0}(0) = \frac{1}{2\pi} \int_0^{2\pi} f_0(t, 0, 0) e^{-jnt}\, dt. \tag{54.3}$$

It is immediately evident that this solution, which is independent of the constant of integration, is periodic with period 2π. From the physical point of view it corresponds to forced vibrations.

In view of (54.1) the approximate solution (54.2) is stable, and to be precise: an arbitrary approximation (53.4) whose initial values x, $\frac{dx}{dt}$ are sufficiently small, will tend to the approximate stationary solution (54.2) for small enough ε and with indefinitely increasing t.

We will now establish the same property for the exact solution of (52.1). For this purpose we observe first of all that for given initial values x_0, x_0' the solution of (54.2) may be represented as a power series in ε. We will then have:

$$(54.4)\begin{cases} x(t) = (x_0 \cos \omega t + \frac{x_0'}{\omega} \sin \omega t) + \varepsilon X(t, x_0, x_0', \varepsilon) \\ x'(t) = (x_0' \cos \omega t - x_0 \omega \sin \omega t) + \varepsilon X_t'(t, x_0, x_0', \varepsilon), \end{cases}$$

where $X(t, x_0, x_0', \varepsilon)$ is an analytical function regular for sufficiently small ε. It is clear that (54.4) will be periodic with period 2π, if, and only if, we have:

$$(54.5) \qquad x(2\pi) - x_0 = 0, \; x'(2\pi) - x_0' = 0.$$

From (54.5) we obtain the following relations for x_0, x_0':

$$(54.6)\begin{cases} x_0(\cos 2\pi\omega - 1) + \frac{x_0'}{\omega} \sin 2\pi\omega + \varepsilon X(2\pi, x_0, x_0', \varepsilon) = 0 \\ -x_0 \omega \sin 2\pi\omega + x_0'(\cos 2\pi\omega - 1) + \varepsilon X_t'(2\pi, x_0, x_0', \varepsilon) = 0. \end{cases}$$

For $\varepsilon = 0$ these relations have the trivial solution $x_0 = 0$, $x_0' = 0$, with a non-zero jacobian

$$(54.7)\qquad \begin{vmatrix} (\cos 2\pi\omega - 1), & \frac{1}{\omega}\sin 2\pi\omega \\ -\omega \sin 2\pi\omega, & (\cos 2\pi\omega - 1) \end{vmatrix} = (\cos 2\pi\omega - 1)^2 + \sin^2 2\pi\omega \neq 0.$$

From this we may conclude that (54.6) has an analytical solution for ε sufficiently small. Substituting this solution in (54.4) we obtain an analytical expression for the periodic solution of the differential equation (52.1). Evidently the constant term in the expansion of this periodic solution is equal to 0. An elementary computation yields for the next term the expression (54.2).

If we continue with the same reasoning which is used in the well known method of Poincaré - Liapounoff, we will readily see that the periodic solution under consideration is stable. For the characteristic exponents are in fact $F'(0) \pm j$ and owing to (54.1), and since ε is always assumed positive, their real part is negative. This proves stability.

55. The second case which we shall now treat is where the equation (53.5) has a non-zero root which satisfies (53.6) and where furthermore f does not contain explicitly the variable t, that is to say where

$$(55.1)\qquad f(t, x, \tfrac{dx}{dt}, \varepsilon) = \sum \varepsilon^n f_n(x, \tfrac{dx}{dt}) .$$

In this case the approximate solution (53.4) assumes the form:

$$(55.2)\qquad x = a \sin \theta + \sum_{m^2 \neq 1} \frac{f_m(a)}{(1-m^2)\omega^2} e^{jm\theta} ,$$

where

$$(55.3)\quad f_m(a) = \frac{1}{2\pi}\int_0^{2\pi} f_o(a \sin \theta,\ a\omega \cos \theta)e^{-jm\theta}d\theta.$$

Thus the approximate stationary solution is of the form

$$(55.4)\quad x = a^*\sin(\nu t+\phi) + \varepsilon \sum_{m^2\neq 1} \frac{f_m(a^*)}{(1-m^2)\omega^2} e^{jm(\nu t+\phi)},$$

where ϕ is an arbitrary constant of integration and

$$(55.5)\qquad \nu = \omega - \frac{\varepsilon}{\omega a^*}\Phi(a^*).$$

It is a ready consequence of (53.1) that for indefinitely increasing t, every approximate solution (55.3) with initial values near enough to those of (55.4) tends to one of the approximate solutions.

It is also immediately clear from (55.4) that in the case under consideration the approximate stationary solution will be periodic with a certain period $\frac{2\pi}{\nu}$, and physically speaking corresponds to free non-dissipating oscillations.

To establish analagous properties for the exact solutions, we observe first that since here the functions R, S, do not contain explicitly the variable t, (52.10) may be written in the form

$$(55.6)\qquad \frac{da}{d\theta} = \frac{\varepsilon F(a) + \varepsilon^2\omega R(a,\theta,\varepsilon)}{\omega^2 - \frac{\varepsilon}{a}\Phi(a) + \varepsilon^2\omega S(a,\theta,\varepsilon)}.$$

Since

$$F(a^*) = 0,\ f'(a^*)<0,$$

it follows from the theorem of Poincaré - Liapounoff that (55.6) has a periodic solution

$$(55.7) \qquad a = \Pi(\theta, \varepsilon), \quad \Pi(\theta, 0) = a^*,$$

where $\Pi(\theta, \varepsilon)$ is an analytical function of ε, regular in the vicinity of 0, and with period 2π with respect to θ.

On the other hand we have from (52.10)

$$\frac{d\theta}{dt} = \omega - \varepsilon\frac{\Phi(a)}{a} + \varepsilon^2 S(a, \theta, \varepsilon)$$

and hence:

$$\{\omega^{-1} + \varepsilon\Pi^*(\theta, \varepsilon)\}\, d\theta = dt, \quad \Pi^* = \frac{\varepsilon S - \frac{\Phi}{a}}{\omega(\omega - \frac{\Phi}{a} + \varepsilon^2 S)}.$$

By integrating we find

$$(55.8) \quad (\omega^{-1} + \varepsilon M_0(\varepsilon))\theta + \varepsilon\sum_{m\neq 0} M_n(\varepsilon)\frac{e^{jn\theta}}{jn} = t + \zeta, \quad \zeta = \text{const.}$$

where

$$M_n(\varepsilon) = \frac{1}{2\pi}\int_0^{2\pi}\Pi^*(\theta, \varepsilon)\, e^{-jn\theta} d\theta.$$

If we set

$$(55.9) \qquad \frac{\omega}{\omega + \varepsilon\omega M_0(\varepsilon)} = \nu, \quad \nu = \psi,$$

then by the implicit function theorem, the solution of (55.8) may be put in the form:

$$(55.10) \qquad \theta = (\nu t + \psi) + \Omega(\nu t + \psi, \varepsilon),$$

where $\Omega(\theta, \varepsilon)$ is an analytical function, regular for sufficiently small ε, and with period 2π with respect to θ.

Thus in view of formulas (52.3),(52.8), (55.7), (55.10), we may conclude that in the case under consideration there is an analytical periodic solution:

$$x = z(\nu t + \psi, \varepsilon),$$

where ψ is an arbitrary constant, ν and $z(\theta, \varepsilon)$ analytical functions regular for ε sufficiently small, and in addition z is periodic in θ with period 2π. If z and ν are expanded in power series and terms of order higher than one neglected, we obtain again (55.4), (55.5).

If we write down the variation equation corresponding to the periodic solution (55.10), it is easily seen that one of the characteristic exponents is 0, while the first term in the expansion of the real part of the other is $F'(a^*)$, hence for ε small enough the real part is negative. By reference to the theories of Poincaré - Liapounoff, we see then that any solution of the differential equation (52.1) whose initial values are sufficiently near to those of the solution (55.10) may be represented in the form

$$x = z(\nu t + \psi, ce^{\rho t}, \varepsilon) \tag{55.11}$$

where $z(\theta, h, \varepsilon)$ is an analytical function of h near $h = 0$, where furthermore $z(\theta, 0, \varepsilon)$ equal $z(\theta, \varepsilon)$ and finally ρ is a characteristic exponent. As for ψ, c, they are constants of integration, with c sufficiently small.

It is thus clear that for indefinitely increasing t, the general solution (55.11) tends to the periodic solution (55.10).

<u>BIBLIOGRAPHY OF N. KRYLOFF AND N. BOGOLIUBOFF</u>

1. Quelques exemples d'oscillations non linéaires. Comptes rendus des séances de l'Académie des Sciences de Paris, t. 194, p. 957-960 (1932).

2. Sur le phénomène de l'entrainement en radiotechnique. Ibid., t. 194, p. 1064-1066 (1932).

3. Les phénomènes de demultiplication de fréquence en radiotechnique. Ibid., t. 194, p. 1119-1122 (1932).

4. Sur quelques propriétés générales des résonances dans la mécanique non linéaire. Ibid., t. 197, p. 908-910 (1933).

5. Problèmes fondamentaux de la mécanique non linéaire. Revue Générale des Sciences t. 44, p. 9-19 (1933). A translation of No. 9.

6. Recherches sur la stabilité dynamique des machines synchrones. (Monographie en russe avec un résumé en français). 100 pp. Kieff, 1932.

7. Recherches sur la stabilité longitudinale des avions. (Monographie en russe avec un résumé en français). 60 pp. Kieff, 1932.

8. Recherches sur la stabilité statique et la stabilité dynamique des machines synchrones. Rapport No. 14 à la 3-ième Section du Congrès International d'Electricité, Paris, 1932. V. 4, p. 179-205.

9. Problèmes fondamentaux de la mécanique non linéaire. (en russe). Bulletin de l'Académie des Sciences de l'URSS, 1933, p. 475-498.

10. Fundamental problems of the non-linear mechanics. Congrès International des Mathématiciens, Zurich, 1932. v. 2, p. 270-272.

11. Méthodes nouvelles de la mécanique non linéaire dans leur application à l'étude du fonctionnement de l'oscillateur à lampe. Partie première. Étude des régimes stationnaires dans le cas de l'absence des forces extérieures périodiques (en russe avec une préface en français). 242 pp., 1934, Moscou, Édit. tech.-théor. d'État.

12. Les méthodes symboliques de la mécanique non linéaire dans leur application à l'étude de la résonance dans l'oscillateur. (en russe). Bull. de l'Académie des Sciences de l'URSS, 1934, p. 7-34.

13. Ueber einige Methoden der nicht linearen Mechanik in ihren Anwendungen zur Theorie der nicht linearen Resonanz. Schweiz. Bauzeitung, Bd. 103, p. 255-257, p. 267-270 (1934).

14. Sur quelques développments formels en séries dans la mécanique non linéaire (Monographie en ukrainien avec un résumé en français). 99 pp., Kieff, 1934. Ukrainska akad. nauk, Inst. mécanique, Rapport No. 5.

15. L'application des méthodes de la mécanique non linéaire à la théorie des perturbations des systèmes canoniques (Monographie en français). 56 pp., Kieff, 1934. Ukrainska akad. nauk, Inst.

mécanique, Rapport no. 4.

16. Méthodes de la mécanique non linéaire appliquées à l'étude des oscillations stationnaires. (Monographie en russe avec un résumé en français), 112 pp., Kieff, 1934. Ukrainska akad. nauk, Inst. mécanique, Rapport No. 8.

17. Sur les solutions quasi-périodiques des équations de la mécanique non linéaire. Comptes rendus des séances de l'Académie des Sciences de Paris, t. 199, p. 1592-1593 (1934).

18. Sur l'étude du cas de resonance dans les problèmes de la mécanique non linéaire. Ibid., t. 200, p. 113-115 (1935).

19. Méthodes de la mécanique non linéaire appliquées à la theorie des oscillations stationaires. Časopis pro pest. matem. 64:107-115. (1935).

20. Méthodes approchées de la mécanique non linéaire dans leur application à l'étude de la perturbation des mouvements périodiques et de divers phénomènes de résonance s'y rapportant. (Monographie en français), 114 pp., Kieff, 1935. Ukrainska akad. nauk Inst. mécanique, Rapport No. 14.

21. New methods in non-linear mechanics as applied to the longitudinal stability of airplanes. Contributions to the Aerodynamic conference, Moscow, 1935. (In Russian).

22. Calculation by methods of non-linear mechanics of the vibrations in lattice girders with consideration

of the normal forces. Ukrainian scientific-research institute of armament, Recueil. Kiev, 1935. (In Russian).

23. Investigation of the influence of resonance in transverse vibration of rods caused by periodic normal forces at one end. Ukrainian scientific-research institute of armament, Recueil. Kiev, 1935. (In Russian).

24. Upon some new results in the domain of non-linear mechanics. Indian Academy of Sciences, Proc. Ser. A, v. 3, p. 523-526. (1936).

25. Application de la mécanique non linéaire à quelques problèmes de la radiotechnique moderne. Onde Électrique, t. 15, p. 508-531 (1936).

26. Sur quelques théorèmes de la théorie générale de la mesure. Comptes rendus de l'Acad. des Sciences de Paris, t.201, p. 1002-1003 (1935).

27. Les mesures invariantes et la transitivité. Ibid. t. 201, p. 1454-1456 (1935).

28. Les mouvements stationnaires généraux dans les systèmes dynamiques de la mécanique non linéaire. Ibid. t. 202, p. 200-201 (1936).

29. La théorie générale de la mesure dans son application à l'étude des systèmes dynamiques de la mécanique non linéaire. Annals of Mathematics, v. 38, p. 65-113 (1937).

30. Sur les propriétés ergodiques de l'équation de Smoluchovsky. Soc. Math. de France, Bull. t.64, p. 49-56 (1936).

31. Les mesures invariantes et transitives dans la mécanique non linéaires. Matematisk sbornik, n. s., t.1, p. 707-710 (1936).

32. Introduction à la mécanique non-linéaire: les méthodes approchées et asymptotiques. Ukrainska akad. nauk. Inst. de la mécanique, Chaire de phys. math. Annales, t. 1-2 (1937).

33. Sur les travaux de la Chaire de la physique mathématique dans la domaine de la mécanique non linéaire. Ibid. t. 3, p. 29-53 (1937) Same in Ukrainian, p. 5-28.

34. La théorie générale de la mesure dans la mécanique non linéaire. (Ukrainian). Ibid, t. 3, p. 55-112 (1937) French résumé, p. 113-118.

35. L'effet de la variation statistique des paramètres sur le mouvement des systèmes dynamiques conservatifs pendant des périodes de temps suffisamment longues. Ibid., t. 3, p. 136-153 (1937). Same in Ukrainian, p. 119-136.

36. L'effet de la variation statistique des paramètres sur les propriétés ergodiques des systèmes dynamiques non conservatifs. Ibid., t.3, p. 172-190 (1937). Same in Ukrainian, p. 154-171.

37. Sur les itérations répétées avec les paramètres variables. Ibid., t. 3, p. 201-211 (1937). Same in Ukrainian, p. 191-200.

38. Sur les équations de Focker-Planck déduites dans la théorie des perturbations à l'aide d'une méthode basée sur les propriétés spectrales de l'hamiltonien perturbateur. Ibid, t. 4, p. 81-157. (1939). Same in Ukrainian, p. 5-80.

39. Sur quelques problèmes de theorie ergodique de systèmes stochastiques. (Ukrainian, without résumé) Ibid., t. 4, p. 243-287 (1939).

40. BOGOLIUBOFF. Sur quelques propriétés arithmétiques des presque périodes. Ibid., t. 4, p. 195-205 (1939). Same in Ukrainian, p. 185-194.

ERRATA

p. 1, l. 11, J_1, J_2 instead of θ_1, θ_2

p. 9, l. 15 should read:

terms of the form tx (a trigonometric function). In the

p. 25, equation (16.2) should read:

$$\frac{1}{2}\left(\frac{dx}{dt}\right)^2 + U(x) = \text{const.} \tag{16.2}$$

p. 45, equation (28.2), delete the subscript o from the $d\tau^2$

p. 51, equation (32.3), at end of equation insert $= 0$.

p. 53, equation (32.8), $G_1(a)$ instead of $G_n(a)$

p. 56, equation (34.7), $\bar{k}$ instead of k

p. 59. equation (36.4), at end of equation insert dt.

p 60, l. 19, $\bar{k}$ instead of k

p. 65, equation (40.5) should read:

$$m\frac{d^2x}{dt^2} + \lambda\frac{dx}{dt} + kx = f(t), \tag{40.5}$$

p. 75, equation (45.6) should read:

$$f(a_1 \cos \phi_1 + a_2 \cos \phi_2) = \sum A_{mn} \cos(m\phi_1 + m\phi_2). \tag{45.6}$$

p. 77, equation (46.9), the first ω should have subscript 1

p. 85, l. 13 should read:

$$\epsilon f(\alpha t,\ a \sin(\tfrac{r}{s}\alpha t + \phi),\ \tfrac{ar\alpha}{s}\cos(\tfrac{r}{s}\alpha t + \phi))$$